THE QUALITY OF THE ENVIRONMENT

The Quality of the Environment

JAMES L. McCAMY

THE FREE PRESS   NEW YORK
COLLIER-MACMILLAN LIMITED   LONDON

THE FREE PRESS
A Division of Simon & Schuster
1230 Avenue of the Americas
New York, NY 10020

Manufactured in the United States of America

10 9 8 7 6 5 4 3 2 1

Library of Congress Cataloging-In-Publication Data
72—80576

ISBN:0-7432-3633-5

For information regarding special discounts for bulk purchases, please contact Simon &
Schuster Special Sales at 1-800-456-6798 or business@simonandschuster.com

contents

preface

THIS IS a book written by a political scientist about a subject of which many people know more than the author knows. A layman in the natural sciences and engineering, I decided that other laymen, as well, had no general introduction to the physical and biological environment. Laymen in the social sciences had no general introduction to the way the available social instruments might be used to save the environment.

I wrote the book, therefore, for laymen. Whenever possible, I used sources not from the technical libraries, assuming that with the aid of generous colleagues I could have understood them, but from books in the Madison Public Library or books purchased in the trade book department of the University Book Store.

For the last four chapters, on the political process, I have not even tried to cite sources for all the conclusions and seemingly offhand statements I have made. To do so would have been to summarize a lifetime of study, teaching, and practice in the political process itself. To the reader who would like to know more about the findings of political science past and present, I recommend my colleague Austin Ranney's, *The Governing of Men*, Third Edition, 1971 (Holt, Rinehart and Winston, New York), an excellent summary.

My thanks to the members of the Wisconsin Seminar on Quality of the Environment, 1967, for their reading of the first two chapters. For reading some of the later chapters, I thank Reid A. Bryson, meteorologist-climatologist, director of the Institute for Environmental Studies; Charles A. Engman, Jr., civil engineer and Associate Director, Institute for Environmental

Studies; Harold K. Forsen, Professor of Nuclear Engineering; Frederick M. Logan, Professor of Art; and Keith McCamy, geophysicist-seismologist, Lamont-Doherty Geological Observatory. All these helpful friends are at the University of Wisconsin. Keith McCamy, our son, is in New York. Lamont-Doherty is a part of Columbia University.

The author's royalty will be paid to the University of Wisconsin Foundation, to be used for further research and teaching in the care and management of the environment.

James L. McCamy
Madison, Wisconsin

part 1

*The State
of the Environment*

chapter

—— 1 ——

WHAT MAKES
THE GOOD ENVIRONMENT?

»»»»»«««««

MAN LIVES in thin circumstance upon the earth. His usuable atmosphere is about ten miles high, and his usable reach in the oceans is about six miles down. Much of the land surface of the earth is unusable by man under present technology. Yet man in his present course seems determined to exhaust the little that he has to sustain life, especially in the advanced industrial nations and most of all in the United States, where population adds to the pressure of technology on resources and tramples what remains of pleasant glades. All this is now a familiar story. It was not familiar in 1967.

Some of us at The University of Wisconsin, with salaries from the U.S. Public Health Service and the University, spent eight weeks then full time in a summer seminar to consider what might be done to slow down, and perhaps even turn, the present trend in the United States toward what we saw as a bleak and dangerous future. We did this voluntarily. We did it for several reasons.

First, we knew that the subject of environment can never be treated adequately from one specialty at a time. The water chemist knows his part of an environmental question. The sanitary engineer knows his part. Too often the piecemeal study of the environment results in a piecemeal patching of one fault at a time. The water may get cleaned by putting refuse into the air, while the soil's contamination is ignored. And seldom are the aesthetic environment or the environment of the senses studied at all. The first is hard to quantify, and the second is relatively unstaffed and unstudied as a science.

Second, we felt that scholars owed it to their fellow men to warn them of what we knew from our studies. The time had come, in the study of environment, when a scholar could no longer decline to take the initiative in reporting recommendations of what should be done. This subject of environment was now as urgent as nuclear weapons were in 1945, when many physical scientists recognized the danger they had created and took the initiative to warn and to recommend.

Third, we were citizens of a local community which we liked and which we wanted to see remain as a pleasant place to live. We looked around us and saw the place where we had chosen to live deteriorating, and we felt that we should go beyond our task as scholars to put our various discoveries and knowledge to the use of our community and our nation. Any one of us could have chosen any place in the world to live. We could have become contract missionaries to an under-developed country, or gone to work for business or another nation, or gone to work for our own government overseas, but we had chosen to stay in the United States, in the State of Wisconsin, in the city of Madison and its vicinity, and we feared that these places would soon be damaged beyond recall. We decided that we were scholars with a civic duty.

Our names and specialties within our disciplines were:

Charles E. Anderson, Meteorology; Associate Director, Space Science and Engineering Center. Specialty: Use of space vehicles as remote sensors for monitoring quality of environment and changes in quality.

Karl R. Braithwaite, Political Science. Specialty: The way scientists and public officials deal with one another as illustrated by the field of oceanography.

Grant Cottam, Botany; Chairman, University of Wisconsin Arboretum. Specialty: Ecology and changes in ecosystems; the preservation of natural sites for scientific research.

Burton R. Fisher, Sociology. Specialty: Social psychology; group attitudes; the influence of environmental change on human behavior.

John L. Lambert, Environmental Science. Specialty: Impact of man on the natural environment.

Philip H. Lewis, Jr., Chairman, Landscape Architecture. Specialty: Identification of natural and social patterns that enhance diversity in the life of man; planning to preserve and enhance the desirable patterns.

Frederick M. Logan, Art and Art Education. Specialty: Cities and urban design as aesthetic experiences for the people living in them; urban change and its consequences for beauty or ugliness.

James L. McCamy, Political Science. Specialty: The consequences for society of the scientific revolution and the demands made on government; the way government responds in policy and administration.

Edward E. Miller, Physics and Soils. Specialty: The physical environment, notably air, water, and soils, and the system of their interdependence; emphasis for this seminar on water in the environment.

William A. Moy, Industrial Engineering. Specialty: The use of systems analysis in the solution of complex problems; the definition of a problem; cost-effectiveness study of some use of the environment.

John E. Ross, Agricultural Journalism; Associate Director, Institute for Environmental Studies. Specialty: Adoption of new scientific ideas by laymen; approaches to the study of the environment.

William D. Stovall, Professor of Preventive Medicine. Specialty: Uses of the environment and consequences in public health; the adoption of public policy for public health; the administration of public health programs. (Retired Director, State Laboratory of Hygiene, after pioneering many practices in public health.)

William B. Stumpf, Environmental Design. Specialty: The sensory-psychological effects on man of various kinds of environment; design to minimize harmful effects and to enhance beneficial effects.

Michael F. Warlum, Wisconsin Idea Theater. Specialty: The arts available in communities; localized art as in community festivals.

We were fourteen men from thirteen fields of specialization. We were chosen by a faculty Committee on Interdisciplinary Studies which had worked for several years on various subjects, including the environment. No effort was made to have all possible fields represented. Conceivably, any field of academic study can be related to the environment of man, and this seminar had to be held within some bounds.

We were chosen according to only three broad specifications. First, the members should come from the biological, physical, and social sciences and the humanities. Second, the members must be respected as specialists in their home fields. Third, the members must be reasonable and open-minded in their willingness to listen to colleagues from other disciplines.

This book was begun in the Wisconsin seminar. The members told one member to write a report, and he began to write the usual kind of report based on the minutes and papers of the seminar. Very soon he saw that the subject of environment was so much bigger than fourteen men could cover in eight weeks that he, or someone, would have to spend months of digging to get even the elementary facts.

An explosion of interest and information about the environment had started soon after the summer of 1967. Books about the environment began to flood the campuses and libraries. Most of them were more argument, alarm, and proposal than facts, or they were collections of readings that usually added up to nonbooks. The best were technical reports, such as one from the American Chemical Society, which were not likely to reach the lay reader.[1]

Yet laymen, especially students, and specialists too, because specialists may tend to see only their own subjects, needed a broad introduction to what happens in the environment and what social instruments can be used to get the changes wanted. What had started to be the performance of a social duty in a limited site by fourteen scholars turned into an intensive effort by the author, himself a layman in all except the social instruments broadly defined. He tried to write a report for laymen that would be factual and neither cry doom nor devils, for the subject is much too large and much too complex to say now that we shall surely perish or that certain groups are to blame for the threats that alarm us.

The "we" of the first chapter and most of the second chapter refers to the members of the seminar. The "we" of the rest of the book is strictly an editorial "we." When members of the seminar had delivered papers that were illustrative of the broader work, the editorial "we" used portions of them. When facts and proposals for the seminar's locale seemed to be helpful to a reader of the broader work, we tried to use them.

Environment to any individual is local. Only by imagination

can he grasp the environment of a nation and a world. At the same time certain aspects of environment, such as the earth's access to the sun, cannot be discussed in one locale. Illustrations not local are taken from the United States.

Few of the men in the Wisconsin seminar knew one another before they began their eight weeks of work together. At no time was there any reluctance to see that a man from a different discipline might have something new and valuable to add. From the beginning each man talked, not as a representative of his field, but as a scholar who applied his knowledge and his sense of method to the subject that was of concern to all; that is, the quality of the environment. It was not an easy summer. The first big obstacle was encountered within the first ten minutes. How do you define quality? Or, can you define quality?

WHAT IS QUALITY?

The most difficult of blocks in any conversation about environment comes with this effort to get agreement on what is meant by good quality. The definition of quality sets the goals for any treatment of the environment, and there are few precise, mathematical, logical ways of considering goals.

We went through several difficult, intellectually frustrating days of trying to get a satisfactory definition. We ended with a partial but workable one. Here are some of the considerations, briefly paraphrased in the form of statements and counter-statements, about the way to reach a definition of quality.

"The definition depends upon the user: a real estate developer has one view; a conservationist has another."

"But there is a public interest that overrides particular interests and that is the view that should prevail. This public interest is frequently defined and promoted by the experts in society, in both public and private bureaucracy, in universities, in interest groups."

"The desires of users should be taken into account."

"No, the weight should be placed not upon what users desire but upon what they ought to have for the sake of the larger society."

"Quality should be defined by the product. Look at the results in health, education, standard of living, ease of living, happiness. When you find all these of high quality, you have found a good environment."

"This only perpetuates the present best from an assortment of environments and does not define high quality except in terms of the present."

"Let the process of decision define quality. Use Programming–Planning–Budgeting for the allocation of funds for use of the environment. This, as a form of system analysis, will produce the main alternative choices to be made, and the choices selected will represent the best that can be done and therefore the best definition. A usable formula would be:

$$Q = \Sigma\, w_i q_i$$

where: Q = quality of the environment

w_i = weights

q_i = individuals' present views of the past, present, and future quality of the environment

"But that formula does not allow for diversity. A better formula would be . . ." and another equation was written in black marker ink on a display scratch pad.

"All right, go into more detail in the first formula. This provides for built-in diversity because the environment is constantly changing and, therefore, the weights and factors affecting the definition of quality constantly change." [2]

"Any proposal for use of the environment must be defensible by a better argument than mere preference on the part of the proposer. For example, we are a group of people with different specialties, but we are very much alike in our ideas of what constitutes the good life. Our choices will be influenced by our outlooks, and that is not good enough as firm evidence."

"True. We all agree that proof must be offered in argument for any proposal. The evidence can be measured for contaminants in the air and water and their effect on organisms and plants. It can be measured for some of the sensory assaults such as noise, and perhaps it can be measured for some of the other factors, such as congestion versus space, horizontal versus vertical forms, trees and grass versus concrete, lakes versus parking lots. There will always be, however, parts of the environment that cannot be measured: for example, the worth of a marsh for scientific research or the worth of a wilderness that is used by only those few people who like to walk and sleep in solitude."

"Once you enter that unmeasurable dimension, your proposals are indefensible except as preferences."

"Not necessarily. Men all the time assign values to unmeasurable things. They do this every time they make a budget that provides for education or defense or public parks. They make these decisions more defensibly if they have a system of weighing alternatives against each other. Hence the new emphasis upon more systematic budgeting."

"Who decides the values to be assigned to the unmeasurable factors?"

"Whoever is making the decision—legislators, board members or committee members, scientists, engineers, architects, administrators."

There was more of such talk. It was clear that quality could not be considered apart from the process by which quality is defined. And it was clear that for those decisions which could not be proved by strict application of scientific methods, the decision-makers who happened to be in the driver's seat made the choice of which road to take. Their choice would affect uncounted generations to follow. The main hope—that their choice would benefit and not damage the future—lay in the effort to see as many alternatives as possible in each decision, to use scientific methods everywhere possible, and to choose those alternatives that were most likely to secure benefits.

The Wisconsin Seminar came out of the exercise with some consensus about quality of the environment, partly for the process of analysis and partly for the desirable attributes of the good environment.

For the process, or how to analyze the environment in order to improve it, we agreed on these points:

1. Evolution, change, is constant and inevitable; therefore, the purpose of concern is not the preservation of what is but rather the maximum social utilization of the results of change and, if

possible, the social choice of various alternatives offered in change.

2. While much change is unstable and does not follow a clean curve, more change is stable and its rate can be plotted. This is especially true for those main elements of change that affect the environment, such as human and animal population, spread of pavement, or increase in automobiles (although few traffic planners seem to have noticed this last truth). Therefore, the student of the environment should (a) ascertain steady trends in change, (b) try to ascertain rates of change in any aspect of the environment, (c) concentrate first on those aspects where the rate of change is accelerating, and (d) try to fix upon the rate of change that will preserve the greatest possible number of alternatives for use in the future and then encourage practices that will be within that desirable rate of change.

3. Some time constants and some cycles are present in any change. These should be analyzed with their rates of change to provide the most clarity of insight. History should be considered along with present change in considering the aesthetic environment. Man's taste and pleasure were fixed in the past as much as they are modified by what he sees and does now.

4. The great handicap to careful use of the environment is the social practice of incremental decision, of doing what is indicated by the last thing that happened. The recommenders of change should try to anticipate the future, so far as the data allows, and should work for change while its direction is still unfixed by preceding events.

5. The student of environment must work with the tools at hand in his time. He cannot anticipate evolution in terms of millennia, nor can he predict far-ahead changes in technology, except by guessing or relying on probability. Despite his lack of omniscience, the student can do more than is being done to define change in the near future. He can project the consequences of contemporary events. He can isolate those events that will have the greatest effect for change in the environment. He can collect events into aggregates and look at a system instead of a singularity. However, with the present tools available and, more acutely, with the limitations of the human mind to visualize and encom-

pass complexity, he should include as few traits as possible within an aggregate for the analysis of a system.

It was easier to talk about the process of looking at the environment than to agree on what traits were desirable in the environment. It would have been easy to state our own biases. As we got to know one another, we turned out to be very much alike in our tastes and tolerances, as should be expected of a group of successful professors in a well-respected university. We were upper-educated, upper-income, traveled, trained to intellectual discipline, kind to one another, and filled with good will toward the whole human race. Nature we loved, and next to nature, art; we "warmed both hands before the fire of life," in the words of Walter Savage Landor, and we wanted everybody to have as good a life as we had. A world designed by us would be a fine world.

The trouble was that our intellectual discipline reminded us that we were not alone and our taste was no defense of a public proposal. The search for attributes of the environment for which we could find scientific-logical-intellectual proof was hard work. It did produce four traits that we thought to be essential to the good environment in any age, for reasons of both science and humanistic values. The traits are all related.

For one, the environment must offer the maximum possible in options of choice at any point in time. This means that an individual or a society will have at any point in time an opportunity to choose the next steps in the use of an environment. Any aspect of the environment that is finished, closed, discarded, whether it be the passenger pigeon or a green field, is for its small part of the universe an end to the options for use of that asset. In exercising the options of choice, men will use up their options inevitably—to kill off the passenger pigeon seemed a good idea at the time and some green fields have to be covered by pavement—but men can calculate more than we do now the destruction of options for the future and can use more sensible economy.

For another, the environment must offer the minimum possible of irreversible traits, and each decision made about its use has to avoid to the utmost the consequence of irreversibility.

This is close kin to the first trait, but it is different. Conceivably the passenger pigeon could be recreated by breeding, and a parking lot can be reconverted to a green field, although neither miracle is probable. Many of the decisions that remove options, in other words, are not irreversible.

But some decisions, and the incremental accumulation of such decisions, create irreversible conditions. When a marsh is dried up and stays dry for a number of years, it will not come back. When that marsh is the saturation basin to retain nutrients that otherwise would enter a lake, the lake is damaged when the marsh dries up. Natural resources of all kinds can be lost irreversibly. Or when a whole society's way of life becomes dependent on the automobile, it cannot revert to the horse. Nature can rebound with remarkable resilience, if allowed to do so; but there are limits even here, and some things of nature can be killed. Society is much less resilient than nature. The social creations easily become irreversible.

For another trait, the environment must offer the maximum diversity possible. We think there is enough evidence and philosophy in the history of man to conclude that diversity is one of the essentials of life, liberty, and the pursuit of happiness, in that most useful and comprehensive phrase from John Locke and Thomas Jefferson.

Diversity is essential in the genetic pool to keep natural selection on the path of survival for the species. Diversity is essential to give the big-brained human a choice of opportunities in the exercise of liberty. Without non-happiness, or without unhappiness, there could be no sensation of happiness. Diversity is essential to provide the opposites. This concept of diversity came up often in our discussion and it came up in relation to many different aspects of environment. It is essential for physical and emotional health; it is essential to a sense of fulfillment and dignity; it is essential for freedom of choice; it is essential for aesthetic pleasure. We concluded that the more monotonous an environment is, the more damaging to the people who live in it; and the more diversity it offers, the more helpful it is to its inhabitants.

Finally, though not in an order of ranking but only as the last point to be entered here, the good environment will allow men to grow in mind and spirit as well as it allows them to stay in good physical health. We make the point mainly because we think it has to be made. So much of the talk about environment is confined to chemicals in the air and water and ingested in food and drugs that we want to make the point that the social, aesthetic, and spiritual qualities are just as important to the life of man. The experienced practitioners of medicine and social science already know that social, aesthetic, and spiritual qualities cannot be separated from physical health, but few have had a chance to talk this way in public, because the approach to environment has been through the narrow tubes of specialization. We did not have time to go deeply into a discussion of social environment, a large subject, and we regret this. But we agreed that the natural environment was intimately related to social, aesthetic, and spiritual health and that it can be just as important in this respect as the social environment.

THE SITE

It is easy to talk about environment in generalities, especially if one mistakes prejudices for evidence. Some of us had read too much on the subject, especially reports from governmental commissions, and had become annoyed that we got no tangible answers but rather only conclusions based on opinions. We decided to study, first, the United States, and second, the county in which we lived plus the county next door. We chose the United States because it represents the industrial world of accelerating technical change with all the complications of urbanism, industrial production that lives by exhausting natural resources, and a complexity of making decisions that may in itself be a threat to the nation's welfare.

The United States represents Western Europe, the Soviet Union, Japan, and those spots such as Sao Paulo or Bombay where industrialism has reached its present stage of advance. The under-industrialized parts of the world have problems of environment, but they are not the same. If we had tried to think about the whole world, we, too, would have floated into vapors of generalities and we would have reached the end of summer with little more than a cry of alarm, a warning of disaster, and a declaration of righteous intent.

We chose the two counties at hand because we already knew something about them and because we felt that they were enough like the whole of the United States to teach us something about the nation.[3] Dane County is the statistical metropolitan area of Madison, Wisconsin. It now has an estimated population of 257,000, and by 2000 A.D. it may have a population of 588,-000. Iowa County, to the west of Dane, is a rural county with an estimated population of 19,500 and a projected population of 17,000 in 1980.

Dane is not Megalopolis and may be located far enough west to avoid being absorbed in the Lake Michigan megalopolis now filling in from Green Bay, Wisconsin, to Gary, Indiana. Nevertheless, Dane County is a metropolis that has, on a smaller scale, the same traits that create a Megalopolis. It has satellite cities that are growing as commuter dormitories. It has traffic problems that call for solution in more than one city at a time. It has overlapping governmental jurisdictions and the usual waste of effort and skill owing to diffused authority.

Iowa County has a rare chance to remain a rural county with all the amenities of country and small-town life. It needs more income. If it looks ahead, it can become prosperous from the money that, according to all forecasts and the investment of smart money, will be spent in large volume for the chance to catch a breath of outdoors in the country.

Whatever has happened, or is happening, in these two counties, one urban and one rural, has happened, is happening, or will happen anywhere in the United States. The nation is now so unified and standardized by the technological revolution that any part can represent the whole. This is our premise in choosing particular sites to illustrate the generality of environment in the United States.

THE QUESTION OF POPULATION

The subject of the rate of growth of population is bound to be raised in any discussion of environment. The very basis of all other concern is the relation of resources to people. When the number of people rises faster than the resources can be renewed or opened, demand becomes higher than supply. Some of the most virulent threats, notably pollution of the air and municipal pollution of water by sewage, are the direct products of a growing number of people. Questions of the quality of the environment reduce in most cases to questions of the impact of man on the environment.

The relation of people to environment underlies all aspects of any discussion. What to do about population is another matter, and the answer is by no means clear. We resolved our discussion of birth rate with such an easy generality that it will not satisfy any reader. There is at this moment in our history no answer in the United States about where fertility rates will go and what future population will be. There are projections, but these may differ from ultimate truth, as any statistician knows.

In an advanced nation, where the techniques of contraception are widely known, the decision of how many children enter the population is made in the privacy of marriage.[4] It is made mostly by white middle-class parents, for we are mainly a white middle-class nation, and the most malicious libel one hears is that a high birthrate is traceable to births to brown, red, and black Americans who are on welfare.

There is almost no way for a civilized society to regulate the intimate decisions between parents, except self-regulation. Some legal steps are discussed. The income tax could penalize a couple with children, in the reverse of its present favor for dependents. Abortion could be legalized, but, while the lowest fertility rate in the world now is in those countries with legalized abortion, there is no assurance that American women would use abortion enough to depress the birth rate.[5] Fees could be charged for each child who attends public schools. To mention these possibilities is to say that, with the possible exception of legalized abortion, none of them has much chance of adoption in a free country with the traditions of the United States.

Social and economic conditions in an advanced industrial nation—a higher standard of living—seem clearly to be more influential in determining fertility rates than access to contraceptive devices.[6] The pill was first licensed in 1960 for general use.

It changed the pattern of fertility in America within ten years. But the trend in the American birthrate had been downward since 1957, so a change had already begun before the pill was available, and again human choice of how many children to have was more decisive than the technology available.

A good natural scientist, trained to look at facts and to add them, looks at the destruction of resources, the chemicals that flood the wrong places, and sees that most of the trouble comes from people. He projects some curves and they all go upward, many of them at an exponential rate, both the curves of destruction and the curves of population. The answer is clear: keep the population from growing so much.

A good social scientist, trained to look at facts and to add them, asks, "How?"

In the early stage of this conversation, the natural scientist gave a quick answer: "Teach people to use birth control; set up clinics to show them how; make the pill or the devices so cheap that all can afford them."

"But, they already know," replied the social scientist. He cited data from survey research to show that only 3 percent of all women in the United States had not heard of the pill in 1965, only five years after the pill was licensed for general use. The most ignorance was among Southern Negroes; 14 percent had not heard of the pill. But, for Negroes elsewhere in the nation, the proportion of Negroes who had not heard of the pill was essentially the same as for all whites, or 3 percent.[7] To know about the pill is not necessarily to use it. Choice still controls, and to some extent economics, although again the decision of how to spend money is also a statement of choice.

"Then teach people the consequences of their choosing to have so many children," said the natural scientist. "Teach them that people cause congestion, damage the natural scenery, destroy places that we may need desperately in the future. Teach them that so long as we spend all our time adjusting to the demands of a larger population, just keeping up with the damage done by too many people, we cannot get started on the bigger job of creating the good society."

"Well, perhaps," said the social scientist. "We can't say

whether education to avoid children for the sake of society would 'take' until we study the responses that we would get. All the evidence indicates that a rise in the standard of living of the American poor would reduce their fertility rate. But, the curve of growth in population would not be changed drastically if the only thing we do is educate the poor. The poor are not the only people who have the children now crowding the schools and landscape.

"Before you conclude that crowds of people prevent the good life in an advanced nation, however, you'd better ask the Dutch. They have one of the highest densities of population in the world, 934 persons per square mile, and they don't think they are ruined. The answer may be in accepting more people but arranging to fit them into the good life by planning how resources should be used." [8]

So it went. Out of the discussion came these instructions to the draftsman of a report. Point out the significance of population to all aspects of the environment. Make it clear that all should be aware of trends in the population and the demands these trends will make on the environment. Suggest that relocating population should be considered to see whether a different distribution of people would place less strain on the environment. Make sure everything possible is being done to keep food production up to demands for food. Stop waste; breed food products for more nutrition; open new sources of food because the United States has to feed not only its own growing population but has to help feed others in the world, especially in times of episodic hunger such as famines and other serious dislocations.

Then we turned to our own sites and made a recommendation for the place we live. Aim for a selective growth of population in Dane and Iowa Counties. Don't strive to attract industry just for the sake of growth. Get the kinds of industry that will cause the least damage to the environment. These are also the kinds of industry that will bring in the high-educated, high-income, professional and white-collar workers who work in offices and research laboratories. These enterprises do not cover the city with smoke or dirty the streams with chemicals. Their workers admire nature and, next to nature, art; and they will be better aware of how the environment should be used as well, and feel more responsibility for future generations.

These are, of course, self-centered solutions, and based on our own bias. The resident of Gary, who is already stuck with the dirty steel industry, will not applaud them.

There is, however, some justification for our parochial concern. We still have a chance to make some choices in our two counties. And the same economic self-interest that welcomed steel mills to Gary moves us to seek the kind of industry that will best serve our future.

We have, relative to the blighted cities, a quality of life that attracts the clean industries. We have honest government, striking scenery, good schools, and easy access to one of the great universities of the world. That quality of life is already being damaged by the growth of population, the damage appearing, as usual, for example, in automobile congestion that gets worse as the conventional treatments by traffic engineers are applied, and in ugly growths of cheap buildings. The sensible solution for our city is to seek those industries that are clean and employ relatively few highly educated people at high pay, and thus depend for our economic growth not upon large numbers but upon high purposes in life. There are industries that our city should shun. There are some others that our city should seek. The growth of population can be selective growth.

If this sounds like a snobbish proposal in terms of the kind of people we think the city should attract, it is not meant to be. The solution was reached in an earnest and humble mood. The evidence and our judgment brought us logically to this end. If we cannot control the fertility rate in America, we can try to control the growth rate by migration to our spot in America, and we can strive to select the kind of people who will make best use of the environment for the future.

THE PURPOSE OF CONCERN

It may be unnecessary, even foolish, to try to find the one central purpose in a concern for the quality of the environment. There are many strands in this matrix of environment. Each one can be significant. As scholars, nevertheless, we feel easier if we can think in terms of the principal goal. We prefer a main goal in which all smaller goals can be seen as steps that lead to it.

In this case of concern for the quality of the environment, the main goal cannot be physical. We cannot predict the cultural terms any more than we can define it in physical terms.

But there is one concept that stands out for the larger purpose. It is the concept of public health. All decisions about the use of environment can be made in the way that will most enhance the public health. This is no place to get into the argument that health has not been defined. We do not attempt to define it, but only to accept it as a concept. Certainly health is affected by the environment. The more perceptive medical men will agree that all illness—which some will accept as the opposite of health— stems from environmental factors—from infection, from allergies, from toxins, from fatigue, unhappiness, inter-personal stress, worry, and the countless other factors that assault feeble man. In this sense, we see the public health as the direct result of conditions in the environment.

Public health, in this sense, is the cumulation of the health of individuals. It includes mental and emotional as well as functional health. Happiness or sadness are as much parts of health as the common cold and insecticide poisoning.

The highest quality environment is one that produces the highest level of public health as measured by the generations alive at any time in history. Proceeding from this concept, we can say that a thousand years from now the environment will be in good shape if it produces the maximum level of public health as defined in that time. And now we can analyze the environment of our time to see if it secures optimum public health.

NOTES

1. *Cleaning Our Environment, The Chemical Basis for Action*, a report by the Subcommittee on Environmental Improvement, Committee on Chemistry and Public Affairs, American Chemical Society (Published by the Society, Washington, D.C., 1969).

2. We did not come out of the seminar with anything so simple as an equation to solve the problem of quality, but we tried. The question is much too complex. Pity.

3. We are carefully avoiding the use of that much misused word *model*. We did not test these two counties to see how much they might be a model, and we doubt that any place can be a true model of any other place. In terms of method, the closest might be *analog*. We really were not trying to be scientifically precise, however, because how could we be when dealing with *quality* where some factors, to be treated at all quantitatively, would have to be assigned weights by the judgment of one group of men?

4. Norman B. Ryder and Charles F. Westoff, "Use of Oral Contraception in the United States, 1965," *Science*, Vol. 153, No. 3741, pp. 1199–1205, Sept. 9, 1966; and "United States: Methods of Fertility Control, 1955, 1960, and 1965," *Studies in Family Planning*, No. 17, pp. 1–5. Feb. 1967.

5. Norman B. Ryder, "The Character of Modern Fertility," *The Annals*, Vol. 369, pp. 26–36, Jan. 1967. The nations are Japan, the Soviet Union, and all countries in Eastern Europe except East Germany and Albania.

6. Kingsley Davis, "Population," in "Technology and Economic Development," *Scientific American*, Vol. 209, No. 3, pp. 63–71, Sept. 1963.

7. Ryder and Westoff, *Science, op. cit.*

8. Of course, densities of population by nations have comparative meaning only when the type of land and resources are also included in the analysis. They are interesting only to show that a dense population can be accommodated in an economy that supports it by a combination of production and trade. Some other densities for Western Europe by number of people per square mile are: Belgium, 796; West Germany, 586; and the United Kingdom, 575. The United States with much greater area, much of it in mountains and desert, has a density of 67 persons per square mile. By 2000 A.D. the estimate is for 99 persons per square mile. U.S. Bureau of the Census, *Statistical Abstract of the United States*, (Government Printing Office, Washington, D.C., 1966), pp. 898–900.

chapter

—— 2 ——

THE LAND

»»»»»»«««««

THE MOST IMPORTANT burdens on the land in terms of effect upon environment are people, domestic animals, vegetation, and solid waste. These can be regarded for our present purpose in the most general way, for again the great value of having a variety of specialists look at a subject is to get away from the microview of any one specialty. A biological scientist thinks of many more forms of life then people, domestic animals, and vegetation. He can examine any one of these forms of life and find a multitude of subdivisions. A citizen acting for a civic purpose, on the other hand, needs to see first the most critical and most immediate influences on the natural environment.

PEOPLE

A baby born in the late 1960's will be in his or her thirties when the century turns. The population of the United States then will probably be 300 million. At the time of its birth in the late 1960's the baby became one of some 200 million people in the United States. The present count is accurate to such an extent that no one needs to worry about using it. The future count, and all others to follow, are projections, and statistical projections are frail and tender probabilities. They are based on curves already established. If the conditions that set the curves change, the projection has to be changed. The safest public policy, nevertheless, is to assume that projections based on present curves are the ones for which all must be prepared.

The new baby of today will be a young man or woman in an even more urbanized society than the one in which he lived his childhood. Today three out of four Americans live in a city. By 2000 A.D. four out of five will live in a city.

This projection does not tell much about the kind of life the young man or woman will lead in urban society. Indeed, there has been too much lumping, and not enough discrimination, about the varieties of urban life in America. Most discussions about the trend of population toward the city is about as meaningless as the Census Bureau's definition of a metropolis as any urban area that contains a nuclear city of at least 50,000 population. A loose meaning of urbanism leads one into the trap of predicting urban population and talking about it as if all cities, and all parts of any one city, are the same.

> The National Planning Association informs us [says *Landscape*] that by 1975, 73 percent of the American population will be living in metropolitan areas. A very impressive statistic until we discover that among these metropolitan areas are Pittsfield, Massachusetts, Waterloo, Iowa, and Laredo, Texas, with a projected population of 88,000.

> At present, 58 percent of our population lives in towns of 50,000 or less; more Americans [live] in towns of 10,000 or less than live in all of the cities of a million or more; and one out of every four Americans lives in a place with less than 2,500 inhabitants. Most of us, in brief, still live in a small city or in a semi-rural setting, and the chances are that even in 1975 the proportion will still be sizeable.[1]

Many of the small towns are near big cities or parts of big cities, but this does not make them metropolitan in mood, as anyone can testify who has observed the small-town, parochial interests of wives and children who spend most of their time in suburbs and have acquired little urban gloss.

Hans Blumenfeld and Gerhard Isenberg have defined the true metropolis, as distinguished from the Census Bureau's metropolitan area, as

> . . . a concentration of at least 500,000 people living within an area in which the traveling time from the outskirts to the center is no more than about 40 minutes. Isenberg and I [Blumenfeld]

have both derived this definition from observations of the transformation of cities into metropolises during the first half of the 20th
Century. At the present time—at least in North America—the
critical mass that distinguishes a metropolis from the traditional
city can be considerably larger—perhaps nearing one million
population.[2]

There is, then, great variety in American urbanism, and we do
not need to think of all the urban population as located in big
metropolises when we contemplate environment. Nor can we
consider all population as being relatively dispersed. There are
all varieties of density: the high density of the center of a city
during working hours; the low density of a small town at night;
the even lower density of a weekend cottage; the high density of
living in the center of a city; the low density of living and working in a small town; the even lower density of living and working on a farm; the medium density of working in a small city;
and the lower density of home and septic tank on one's own little
acre at the edge of a farm. This point of variety in the densities
of American population needs to be made again and again whenever statistics are presented in a way that suggests uniformity.

    Not all great cities are alike, not even in shape. London with
its green belt is as different from the Amsterdam, Rotterdam,
and Utrecht conurbation as New York, with its metropolitan region comprising sixteen counties in three states, is different from
Moscow, continental Europe's largest complex based upon one
city center.[3]

    Not all urban life is the same. Any experienced city dweller
knows that home can be a single-family house in either the oldest
or the newest part of the city. Or, home can be a high-rise building facing a park or river, or in a high-rise building that faces another high-rise building. Home can be a slum room or in an
apartment building with a doorman to control the entrance.
Home can be in a building old or new. Home can be in the center or the outer fringe of the city. Home can be in a concentration of mobile homes crowded into some corner of the center or
the fringe. Home can be a hotel. Home can be a cottage beyond
the last sewer line but near enough to allow a working life in the
city.

For those who do not work in the city, variety exists as well. Home can be a self-owned family farm or a worker's house on a corporate farm. Home can be in a rural slum of huts once used for migrant workers and now become permanent dwellings. Home can be in a passed-over Appalachian community or on a forgotten Indian reservation. Home can be in a rural concentration of mobile homes. Home can be a house with a view or a house that huddles in a coulee against the wind. Home can be a house in a village with or without running water. Home can be in a warm country and without central heat or in a cold country where life is hard without central heat.

Densities of population vary, and all the pleasures and the problems that come with people vary.

Perhaps it is safe to make three generalizations about the future of urbanism, based on the clear precedents that have appeared in cities up to now.

The first is that cities will continue to be grouped around central districts, with variations, of course, for the cities such as Los Angeles that have grown since the automobile became prevalent. Despite the widely held assumption that the Atlantic Seaboard, the Great Lakes, and the West Coast metropolises will merge by overlap, each metropolis in the chains has its own center and its own identification. A man says, "I'm from Newark," not "I'm from the Atlantic Seaboard megalopolis known as Boswash!" [4] The man acts and thinks as if his city has its own center and its own identity. "Conurbation can occur," says Blumenfeld, "only when two expanding centers overlap, and, except perhaps between Washington and Baltimore, that is not likely to happen anywhere in North America during this century." [5]

The second generalization is that the size of a metropolis will not be, perhaps cannot be, confined by law. Efforts to do so have failed in nations with strong authorities. Both Elizabeth I and Oliver Cromwell, strong rulers, tried to limit the size of London by circling it with a green belt. The result was overcrowding and a failure of the plan. In modern times, with the Town and County Planning Act of 1947, a green belt again became a fundamental aim in plans and in law. The metropolis leaped across the green belt and continued to grow.

The difficulty is that there is another, deeper, sort of reality: the economic and social reality of people's jobs and homes, and the way they travel between them. Since 1945 London has continued to grow, and grow rapidly: but because the planners would not let it sprawl, because it has been hemmed in by the Green Belt, it has grown in new ways. In the zone beyond the original 5-mile-wide Green Belt, that is between 20 and 40 miles from central London, the existing towns have swelled; and new towns have grown out of villages, or on virgin fields, into major centres. Altogether, this "Outer Ring" added nearly one million to its population in the decade 1951–61, representing two-fifths of the net growth of the British population. Of course, not all these people look to the Greater London Conurbation for a living; but many—it is not quite certain yet how many—travel across the Green Belt into London's centre or its suburbs, each workday morning, to earn their daily bread. So, in an important sense, towns 20–30 miles out . . . have become parts of London too. Yet since 1945 they have grown in the way they always grew, not a part of a single sprawling urban mass, but as separate entities each with its own individuality: planning has seen to that.[6]

A democracy in modern times, in other words, can plan the growth of a metropolis but cannot stop it.

Nor can a totalitarian state, apparently, confine the growth of a metropolis. Karl Marx condemned big cities because in his day pollution of air and water was most conspicuous in the cities, as it still is now, with congested automobiles added to industrial and household smoke. The Soviet government accordingly tried to limit the size of cities. It failed. Even when the revolution and war halved the population of Leningrad twice, after 1917 and again after 1941, the city regained its numbers and continued to grow. In the late 1960's Leningrad, with a population of 4 million, had four times the population it had in 1921.

The Central Committee of the Communist Party of the U.S.S.R. had, in 1931, already declared itself against the further growth of big cities and from 1932 the policy was to restrict further industrial growth in Moscow and Leningrad. In the 1935 General Plan of Reconstruction for the City of Moscow, limitation of the city's growth was made into a central planning objective. . . . The future population target was to be five million, which the city would reach eventually by natural increase alone; net immigration was to be cut to zero.[7]

The Central Committee can act that way. It controls the real estate. But it could not stop the growth of big cities. Despite the attempt to prohibit net migration into Moscow, the population by 1939 had reached 80 percent of the ultimate limit set by the plan. By 1959 the population had gone beyond the 5 million set by the plan. But Moscow as defined in 1932 was no longer the true city. Suburban areas had grown steadily so that, to face reality, the city's area was doubled by decree in 1960 and a million people were added to the Moscow population to make no more than a memory of the 1935 goal.

The big cities have a disposition for growth. Social and economic necessity made them big in the first place and continues to make them grow. And the firm determination of queen, or dictator, or central committee cannot stop that growth. Only the British scheme for planning a growth, by placing a green belt around the central city, seems to have a chance. This experience, however, is only twenty years old and hardly final as a test.

The third generalization about cities is that the political boundaries of city, county, or in some cases state in the American system of federalism have little relevance to the extent of an urban population. These lines on the land were drawn before the railroad in most parts of the nation, before the hard-surfaced highway and the high-speed automobile, before the elevator and the telephone and all the other modern devices that have changed man's use of the city. When a central city is itself an enormous complex surrounded by other complexes, sometimes in different states, it becomes hopeless to think of the distribution of people according to the archaic limits of political jurisdiction.

Whatever new lines are drawn to help to assure, in Coleman Woodbury's words, "habitats that will be kindly in the lives of men," will vary from one site to another.[8] The best concept to provide the kind of area for adequate treatment of the urban environment is regionalism and regional planning. A region is whatever area is defined as such by common ties among people in economic, social, and traditional relationships and in ties imposed by such natural features as climate, physiography, or types of soil. A region can be a river valley or, for our purposes, it can be an urban system tied to a rural system. The point is not to argue over definitions but to accept the fact that American cities as environment are not confined by the city limits or, in many cases, by county or state boundaries.

THE METROPOLITAN AREA
OF MADISON, WISCONSIN

With planning and control, it is still possible to preserve habitats that will be kindly in most of the metropolitan area of Madison, Wisconsin, to illustrate with a small example. Many other cities in the United States share the same condition.

The metropolitan area, in this case, can be defined as the whole of Dane County. There is no urban overflow into surrounding counties. On the contrary, the population diminishes toward the edges of Dane County as the distance from the central city increases. No other cities approach the size of Madison for miles around, the nearest being Milwaukee whose western urban edges (not the political boundaries but the true urban area) lie about sixty miles to the east of Madison. Travel time from the edge of the county to the center of the central city is no more than forty minutes by automobile under present traffic circumstances. It would be less, of course, with high-speed rail trains.

Madison in this definition of a metropolitan area consists of a central city of about 180,000 population. (It also has the typical political malady of four other municipalities converged with the central city. They are Monona Village, Maple Bluff, Shorewood, and Middleton, all dormitories for people who live there because the central city exists). By 2000 A.D. the metropolitan county is expected to have 588,000 population.

The central city is sprawling at the edges, a condition dictated by the interests of real estate developers. This too is typical in America. The real estate men decide the shape of American cities. City planners and commissions have no authority except to enforce zoning and to insist upon a few elementary rules about streets, parks, and sewers.

Beyond the edge of sprawl come the septic tank suburbs, houses scattered through the open country on the edges of farms. These have met, in Dane County, standards for their own wells

DANE COUNTY POPULATION DENSITY

LEGEND

over 3000 persons per square mile (City of Madison)

100-400 persons per square mile

50-99 persons per square mile

25-49 persons per square mile

less than 25 persons per square mile

⁴⁰ Town or City of Madison Average

0   3   6
Miles

and for their septic tanks and fields. The inhabitants work in the central city.

Next is open country that still, in this county, reveals pleasing variety. It includes dairy farms that alternate fields of corn with fields of pasture. It also includes hog farms and herds of steers to be fattened. It has woodlots, ravines, rolling hills, copses, coulees, lakes, ponds, and meandering streams. It has marshes and some fields that are ponds in the spring but dry for the summer. In winter the land can be snow-covered for weeks at a time. The quality of the soil varies from untillable to rich, and a casual glance by a Sunday driver can usually tell him the quality of the soil by the appearance of the farms. The least painted houses, the smallest barns, and the look of dilapidation go with poor soil and poor terrain for farming. Some farmers never had a chance to prosper.

Scattered in this idyllic countryside are the villages. These were once, before the automobile, centers for rural business. Feed stores, farm machinery, clothing stores, saloons, and the services of lawyers, doctors, and barbers were to be found in each one. Bars (in local usage called taverns) and stores that sell farm supplies and machinery may still be in business. Other stores and services have been changed to supply the new needs of both farmers and village residents who now work in the central city. The doctors and lawyers have mostly disappeared. These villages bear the proud old names of Black Earth, Cross Plains, Verona, Waunakee, Oregon, De Forest, McFarland, Stoughton, Sun Prairie, Cottage Grove, Deerfield, Marshall, Cambridge. They still have shaded streets, quiet nights, and the restful tempo of small towns. The older houses have high ceilings and large rooms. The newer houses look the same as the new houses on the edge of the central sprawl or in the septic tank suburbs. Whatever the age of the house, the inhabitant lives a small-town life after he returns from work in the central city.

These villages have been growing in population since 1940; still, in 1960 the largest, Stoughton, had only 5,555 people.

These new-old villages are ready-made towns within the green belt. When they are seen as part of the whole metropolitan area they appear as dormitories that offer many more amenities than the bleak edges of the central sprawl and they offer the solution for a planned dispersal of the new population to come. The green belt exists; so do the villages within and beyond the green belt.

The projected population of 588,000 by the year 2000 will be much easier to live with if these people are distributed over the county than if they are concentrated. The habitat will be much more "kindly in the lives of men" if groups of people are separated by stretches of green and men have fields and woods to walk in. Dane County, the metropolitan region of Madison, is blessed with more than the usual natural beauty in both landscape and townscape. Its job is less to create a kindly environment than to preserve the one it already has.

RECOMMENDATIONS FOR THE DISTRIBUTION OF PEOPLE

Our thesis has been that it is not so much the number of people in a city as how they are distributed that makes life easier or harder to live. In metropolitan Madison the distribution has already started on a base of the central city plus the revived and growing villages that lie in the green belt. The best way to maintain the amenities of urban life when the population has doubled is to keep the pattern of green countryside mingled with clusters of people within forty minutes of the center of the central city.

1. Zoning should be handled for the whole county as one unit. An approach to the new urbanism from the confines of a city, a village, a township will not make life more livable. In the older cities that have grown in chaos this ancient system has meant nothing but harm to the good life. In our locale the facts indicate county-wide regulation of the distribution of people. (In some other places regional handling will be indicated, sometimes across state lines).

2. The zoning should provide variety in the daily routines of people. This means congestion in the center of the central city as well as green open spaces. There is a pleasure in "going downtown" just as there is pleasure in seeing a herd of cows on a quiet hillside. The main purpose of zoning is to mix green intervals

with built-up intervals, to preserve the heritage of countryside as urbanism grows.

3. Use other incentives than zoning to get people to settle in recommended places. Government has an unlimited variety of incentives as well as sanctions it can employ and has used a great many in the United States. In the management of environment there is no reason why incentives cannot be used. Tax deductions, easy loans, uniform quality of schools, subsidies to transit companies—all these and others can be offered to persuade people to live, let us say, in a village instead of at the sprawling edge of the central city or in the core.

4. In the use of zoning and incentives, the plan should be comprehensive. It should provide for the following, among others. The best soils and terrain for farming should be kept open for farming. Highways should be planned so that traffic flow will preserve and best utilize natural assets. Marshes that are tributaries of lakes and other wet lands should be saved from building developers. Natural scientific areas should be saved. Historical sites should be saved. Always the plan should take account of vistas, variety, ease of transit, and all other traits that make life easy or hard for the human who lives amidst urbanism. Madison–Dane County as an urban area still has so many assets that can be saved while at the same time the population is expanding.

5. Begin now to tie the whole new urban complex together with a rapid transit system, using rail lines already installed wherever possible. Madison can avoid the agonies that now beset New York, San Francisco, Chicago, or Los Angeles. Unless the incentive of easy, accessible fast mass transit is offered, people from the edges of the central city and people from the outlying villages will drive their cars to work, thus creating intolerable congestion and intolerable air pollution. We think it is inevitable that if Madison is to remain a kindly place in which to live, mass transit must be provided and the number and kind of automobiles to be allowed in the center of the central city must be regulated. This will be true even after the internal combustion engine is replaced by a less polluting engine. There is simply not enough room at the center of any city for all the automobiles and all the people to mingle in peace.

The present use of land in the city and county allows many of
the traditional amenities. The following table shows the various
uses of land. This space can absorb new population and still preserve the same relative distribution if a plan is made for the
whole county.

Land Use: Dane County and City of Madison

(1) Land Use in Dane County exclusive of use in Madison.
Percentages are based on total land in the county. The percent-
ages are based on 1964 figures and were obtained from the Dane
County Planning Department.
The total acreage in Dane County is 765,168.3 A.

(2) Land Use in Madison.
Percentages are based on the total land in the city in 1964. The
source is the same as given in (1) above.
The total acreage in the City of Madison is 27,790.0 A. Percentage of County area in Madison = 3.7.

Use of Land	(1) Percentage of County Acres	(2) Percentage of Madison Acres
Residential	2.5	18.8
Commercial	.1	2.2
Industrial and Transportation	.6	7.2
Public Right-of-Way	2.9	14.8
Institutional	.5	8.1
Open Space and Recreational	1.2	10.0
Agricultural and Vacant	88.3	38.9
	96.1	100.0

Total land area includes ponds, lakes, and rivers under 40 A. in area.

Agricultural and vacant includes all vacant land in the county or city
no matter what the zoning is.

Institutional includes government offices, police stations, fire stations,
post offices, and garages; schools of all types; libraries and cultural
centers; historic sites and monuments; hospitals of all types and medical research institutions and convalescent and nursing homes; military bases, forts and camps, missiles and radar; religious buildings;
charitable buildings (except offices); public indoor amusement and
recreation, including armories; cemeteries.

Open space includes reserves, refuges and fisheries, parks and arbore-
tums, public and private golf courses, fairgrounds, stadium and race

tracks, land fill, dumps, burning sites, institutional or experimental farms.

A PLACE TO GO

People who live in cities need an easy way to get away from cities for a change. All over America the least productive land in the country can be saved for recreational use by urbanites.

In the case of metropolitan Dane County-Madison, our escape park is Iowa County next door. Some of Iowa County is good farmland. Much of it is marginal farmland at best, and much of the county is untillable because of terrain. The same terrain gives its scenery exciting variety and makes it an ideal park for driving, picnics, weekend cottages, and pleasant dining-out.

Recreation is a business and a good one for Iowa County to develop. The result for metropolitan Dane County will be a nearby rural haven to which people can go for a change from urban living.

DOMESTIC ANIMALS

Livestock and poultry population is not discussed as much as human population but animals may exceed people in their impact on the environment. The disposal of human waste is relatively simple compared to the disposal of animal waste. Some biologists and agricultural engineers are concerned. They write articles and hold conferences and have built a beginning body of findings.[9]

There are half as many cattle as people in the United States, 112,330,000 in 1970, of which only 13,875,000 were milk cows. There are 20,422,000 sheep and lambs and 56,743,000 hogs and pigs, again using figures for 1970.[10] There are twice as many chickens as people, 432,000,000 in 1969.[11]

Most of these farm animals are kept in small spaces. Beef cattle typically spend one-third of their lives in feedlots to be fattened for market. Most egg-laying hens and most broilers are raised in confinement, and the number of birds in one installation can go as high as 1 million. Swine are already raised in sheds in large numbers, and the trend is toward raising all swine in confinement and in larger numbers for each building. Dairy cattle are increasing in number for the space allowed, automation in feed-ing, milking, and cleaning having made it possible for one man to handle more cows.

Only the dairy cattle, and to some extent the swine that are still not raised in confinement, are likely to be adjacent to grass and crop land in sufficient acreage for the manure to be used economically as fertilizer. The new massing of animals and the new methods of feeding for most livestock in the United States raise problems of waste disposal that are much more serious than those caused by the human population.

Farm animals in the United States

> . . . void over 1000 million cu yd of solid wastes per year. The magnitude of this figure can be better appreciated when compared to the 200 million cu yd of sludge produced from the human pop-ulation of the United States. On the basis of population equiva-lence reported by Taiganides and Hazen (1966) the daily wastes from poultry, swine, and cattle alone are equivalent to 10 times the wastes of the human population of the United States.[12]

There are aesthetic problems that come with manure dust and bad odors. The main result of all these animals in the environ-ment is, however, that surface water runoff carries so much fertil-izer into streams. In slow streams and lakes, plants grow only to die. Organic pollution demands oxygen. The combination of plants and bacteria uses up more oxygen than is available for a good biological balance. Fish begin to die. Rotting algae begins to stink. A health threat from drinking or swimming in contami-nated water is added to the aesthetic discomfort. There is no longer any question that some new method of disposing of ani-mal waste has to be adopted in the public interest.

Unfortunately, the present technology for disposing of animal manure cannot be characterized as very advanced when it has to be used on one farm or one feedlot at a time. It is one thing to treat human sewage that is collected through an elaborate system of pipes in a city. It is another thing to treat animal sewage in scattered sites, something like the haunting thought of providing septic tanks for herds of animals with capacities ten times greater than for humans.

Storage in bulk and use as fertilizer is the simplest method of disposal. But this is not as simple as laymen might suppose. For one thing, manure is not always the cheapest fertilizer, especially when it has to be hauled very far. For another, flies, odors, and dust in storage are ever-present problems. In most parts of the country the use of manure as fertilizer is limited to certain times of the year. Land is prepared in the fall and spring. In many places the ground freezes and will not absorb anything spread on the surface. Snow covers the ground. To spread manure on frozen ground or snow is simply to add it to runoff into streams.

Los Angeles has what must be the largest pile of manure in the world. It is an example of the advanced stage of plain storage. Two hundred dairy farmers owning 50,000 cows belong to the Dairyman's Fertilizer Co-Op, which maintains the pile. Weekly cleanings are added to the pile and three salesmen find customers mainly among citrus and vegetable farmers, nursery firms, and highway landscapers. The selling price does not pay the cost of hauling and storage, but this is still the cheapest method the dairymen can find to dispose of the waste. The pile is 50 feet high, covers four acres, contains 400,000 cubic yards, and it stays about the same size no matter how hard the salesmen work to cut into it.

An alternative to bulk storage is the lagoon, surely one of the most misused words in the English language. Manure is dumped into an uncovered pond, the lagoon, and biological activity is supposed to reduce the manure to sludge. The sludge is then spread as fertilizer. To make this work as well as it is supposed to work requires much more land area for the pond, to allow enough time for bacteria to work, than most farmers can afford. As now used, the pond method does little to reduce solids, and the problem of odors remains as in bulk storage.[13] In cold country, sludge dumped on frozen ground or snow is still a polluter of streams. Whether the manure is in bulk or sludge, it must be stored through the winter months.

The federal government decided to promote bulk storage when it approved the first pilot program in the nation for the same Dane County with which the Wisconsin Seminar was concerned. Under its program of agricultural stabilization and conservation, propaganda euphemism for farm subsidy, it will pay

80 percent of the construction cost of approved facilities for the storage of manure. The upper limit for any farm is $2,500. The first unit approved was to be 36 by 36 feet, to take the manure from 60 cows and hold it for 180 days. It was a simple structure. The concrete wall of a barn was one side; a seven foot concrete wall the other.

Dane County was a good place to begin. Its Lake Mendota is a regional attraction, and it is dying faster than it should because of the nutrients that flow into it from farms and city runoff. While no money was available for any purpose save farm storage of manure, the agricultural officials hoped to get the cooperation of the cities that also contribute to the nutrition of plants in the lake.[14]

VEGETATION

Common as it is, vegetation is also an issue of environment. The question is how to keep enough of it to serve the needs of other forms of life. The big issues are national: what to do about the plowed and eroding plains, how to save wilderness areas, how to save a forest of giant redwoods that are an asset for all mankind. The local issues are just as important, however, and they get much less attention.

Man has had some impact on vegetation ever since plants began to grow in our particular locality.[15] He was in Southern Wisconsin when the glacier melted and plants could begin to grow. Indians periodically set fires to clear the land. As one result, vegetation changed to communities of fire-resistant plants. Europeans introduced large-scale farming and the cutting of timber, and put livestock on the land.

But change is normal and inevitable. Biotic communities are dynamic systems. "They are never the same in space or time. They respond to changes in weather, changes in disturbance, and to many man-induced factors such as the introduction of new species."

Some of the change will be drastic. Immediately after the retreat of the glacier this area was covered by a forest similar to that found today in much of southern Canada. Gradual change brought, by about 2000 B.C., a climate much warmer and dryer than now and a vegetation that was largely grassland with a few oaks. Other changes followed. A second warm and dry period prevailed about 800 years ago, and a colder period came on in the late nineteenth century.

At the time of European settlement, 1830–1850, about 16 percent of the land was grassland and 80 percent was covered by oak savanna, "characterized by a very sparse growth of trees (about one-tenth the number per acre as occur in our present forests) and an undergrowth of the same species that occurred in the grasslands." The remainder of the land was in marshes and lakes.

Then European farmers introduced their drastic changes. Portions of the land were turned into planted crops. The fires were halted with the result that many of the oak openings and some of the prairie turned into forest. The oak savannas that did not turn into forests disappeared. "The old savanna trees are still evident in the forests; some of them remain in land that has been maintained continuously in unimproved pasture; and a few are still evident among the houses in our urban developments. As a community, however, they are gone."

So are the prairies gone, plowed under for crops. "The only remnants of this community are to be found along railroad rights-of-way and similar places, where the absence of grazing and the presence of fire have approximated the conditions prevailing before the coming of white man." Some grassland still exists, under an unusual circumstance such as on a steep hillside with shallow soil where the available moisture is too low for the growth of trees.

"It is apparent that there is practically no land that exists in the condition it was before the coming of the white man. The largest portion of this land has been converted to agriculture. The remainder has been changed because of the cessation of the practices followed by the Indians." But the arrival of the white man was only a different kind of human impact, not the beginning.

Change continues. In 1959 J.T. Curtis was concerned with non-crop vegetation over the whole state as it existed without major disturbance.[16] Since then, in less than ten years, about half of his examples of relatively undisturbed vegetation disappeared.

The urgent need today is to begin to manage the environment so that enough vegetation will remain for photosynthesis, for food, for pleasure, for scientific history and research. Growing populations in growing cities cover with pavement and buildings land that once was covered by vegetation. Real estate men drain wet lands for crops and buildings. Farmers and timbermen disturb forests by cutting and grazing. All man-made change can disturb native shrubs that shelter wildlife until the pleasures of nature are lost to memory.

The key to safe management of the environment for vegetation is the preservation of options.

"In the absence of any clear understanding of what constitutes quality for the present," says Grant Cottam, and with the knowledge that man's needs and wants are likely to change with population growth and with scientific and technological advances, we are likely to be more successful if we define quality in terms of process rather than in absolute characteristics. For those lands that are to be maintained in a relatively natural state, and for many that are not, quality may be related to the ability of the land to maintain itself.

"This involves the prevention of erosion, the prevention of loss of nutrients and organic matter, and the prevention of loss of species. The plants and animals, if undisturbed, reach an equilibrium with each other and with the physical environment. Gradual changes occur over long periods of time, but normally new habitats are created as old habitats change, and there results a mosaic that is relatively stable. Losses in soil and species are usually absolute.

"Stated in a more positive way, the above assumption may be called the preservation of options. Implicit in this assumption is the idea that we are not sufficiently knowledgeable to know what can be safely dispensed with, and that we should maintain portions of all kinds of natural environments as a hedge against future developments, developments that may require products or insights that can be obtained only from natural environments that at the moment are useful for different purposes. . . ."

SOLID WASTE

In some ways, solid waste may be the environmental difficulty that will be easiest to overcome. This waste is physical and will submit to engineering. It has a bad standing in the minds of people, who call it garbage, rubbish, trash. Some of it is already valuable enough to re-use and more of it can become valuable once more research is devoted to the economy of re-use. It is tangible, and people can take such things as bottles and paper to be recycled.

Yet, in another way, solid waste may be as troublesome as the other threats to the environment. Like the federal budget it is so vast and so varied that no one can really comprehend it, so we may tend to let it ride along, depending upon others to do the right thing. Much of it comes from the habits of consumers, who have become used to plastic packages, throw-away bottles, thick newspapers, and vending machines that serve food in discardable cups and packages. Once a college classroom building has installed food-vending machines, a sensitive professor who meets a class in the late afternoon feels that he is not in a hall of academe but in the middle of the town dump. While people may not defend trash as a concept, they can become very defensive about their right to unlimited use of the by-products of our lavish technology.

Much waste comes from industrial processes that began before we recognized the problem to be grave. The processes became both habit and economy and to change them now is seen as a threat to profit.

A citizen can start with quantity. In 1964 each person in the United States produced 150 pounds of garbage, 1000 pounds of combustibles except garbage, and 450 pounds of noncombustibles. He can learn that in 1966 waste in cities was composed of 64 percent rubbish. He can learn that paper as a component of rubbish makes up by far the bulk of waste—42 percent of all municipal solid waste. Food wastes make up 12 percent of the total, and 24 percent is non-combustibles (metals, glass and ceramics, ashes).

He can find each of these main categories sub-divided into an

impressive number of items in which, for example, rubbish includes not only household and industrial trash but street refuse,
dead animals, and abandoned automobiles and trucks.[17]

A citizen can find a more detailed, analytical listing of the
components of solid waste that begins with Acetylene Wastes and
ends with Zirconium, plus methods for recovering useful scrap,
from market quotations for scrap metal. But if he is a layman, he
will have rough going.[18]

If he is of the urban majority, this citizen probably has never
thought of rural waste, but the rural environmentalists are aware
of it.

In addition to animal waste,

> The sources of farm wastes are: (1) Human wastes from America's
> 13 million farm and 26 million nonfarm population . . . (2) Crop
> residues, such as sugarcane, cornstalks, and pea and tomato vines
> amount to 8 tons of plant waste for each American family . . .
> (3) Wastes from rural fruit and vegetable processing units . . . and
> other rural industries . . . (4) Approximately 58 million dead
> birds per year. . . . Even under good management, the average
> mortality rate in hen production is about 1% per month. This
> means the disposal of 1000 birds per month from a well-managed
> 100,000 laying hen production unit. (5) Residues from agricultural
> chemicals such as pesticides, plant nutrients, and control agents.[19]

This listing leaves out abandoned farm machinery and vehicles, dead animals other than chickens, and any of the other categories of solid waste that can be cast off in the country as well as
in the town.

METHODS OF DISPOSAL

The easiest and most obvious way to dispose of solid waste is
to dump it—in the backyard, at a town dump, on land the city
wants to build up, called landfill, or in the ocean, if the city is on
a coast. When city officials say they are running out of places to
put the stuff, they are talking about this kind of dumping. It is
limited by the time the material requires to degrade to its original elements. Moreover, while waiting to degrade, it can pollute
the water and it can stink to heaven unless managed very carefully.

Another method is to compress, grind, shred, flail, or pulp the

waste before dumping it so that small bundles, pellets, or bri-
quettes take up less space when used for landfill and thus postpone the time when places to put waste will be used up.

An interesting variation of both these methods is to go beyond
ordinary landfill and to build scenery. A city can build a hill and
every foot of elevation will represent a foot of dump waste.
While many cities need hills, this method too is finite. When
well-managed, all landfill, flat and elevated, is covered with
earth; and trees, grass, and flowers are planted to make pleasant
new land for whatever purpose the city decides.

Another method is burning. This has the advantage of turning
combustible waste into smoke and ash and, sometimes, of providing heat for other purposes. But burning pollutes the air and occasionally gives off bad odors. Reducing the size of waste by all
the means used to prepare waste for landfill can also be used to
prepare waste for burning.

Salvage is also a method of waste disposal, involving mainly
scrap metals, glass, and paper. Before the useful waste can be sal-
vaged, it must be separated. This can be done by techniques that
vary from hand sorting to chemical processing.

With the exception of salvage for the purpose of recycling, all
these methods simply change the form of the waste into something else that requires time to allow the natural process of oxidation to reduce the amount. At present we are piling so much
waste upon waste that oxidation cannot keep up with it, and cities and farms are desperate to find dumping places. Some cities
try to buy or lease distant land and propose to haul their waste
by train to the dump. People near the dump ground understand-
ably object. The only satisfactory final answer seems to be salvage
and recycling.

NOTES

1. "Notes and Comments," *Landscape*, Autumn 1967, Vol. 17, No. 1,
   p. 1.
2. Hans Blumenfeld, "The Modern Metropolis," *Scientific American*,
   Sept. 1965, Vol. 213, No. 3, p. 64.

3. Peter Hall, *The World Cities*, (McGraw-Hill, New York, 1966), Chaps. 2, 4, 6, 7. "Conurbation" means a metropolitan region that forms from the growing together of neighboring cities.

4. Herman Kahn and Anthony J. Wiener, in an excellent book, were seized with cuteness:

> The United States in the year 2000 will probably see at least three gargantuan "megalopolises" that we have labeled—only half frivolously—"Boswash," "Chipitts," and "Sansan." Boswash refers to the megalopolis that will extend between Boston and Washington . . . (We might even call it "Portport" on the grounds that this megalopolis really stretches from Portland, Maine, to Portsmouth, Virginia). Chipitts . . . may stretch from Chicago to Pittsburgh . . . Sansan would . . . stretch . . . ultimately from San Diego to . . . San Francisco.

The Year 2000, A Framework for Speculation on the Next Thirty-three Years (The Macmillan Co., New York, 1967), p. 61. These terms threatened to get loose in public when *Newsweek* picked them up, Nov. 27, 1967, p. 89.

5. Blumenfeld, *op. cit.*, p. 72.

6. Hall, *op. cit.*, p. 31.

7. Hall, *op. cit.*, pp. 158–61.

8. Coleman Woodbury, "The Role of the Regional Planner in Preserving Habitats and Scenic Values," in F. Fraser Darling and John P. Milton (eds.), *Future Environments of North America* (The Natural History Press, Garden City, New York, 1966), pp. 568–70.

9. For example, see *Management of Farm Animal Wastes*, proceedings National Symposium on Animal Waste Management, 1966. American Society of Agricultural Engineers Publication No. SP-0366.

10. U.S. Bureau of the Census, *Statistical Abstract of the United States, 1970;* Table No. 975, p. 618, figures for Jan. 1, 1970. Oddly the Census does not report cats and dogs. A manufacturer of pet food estimates there are 38 million cats and about the same number of dogs. *The Wall Street Journal*, Nov. 27, 1970. Dogs became a cause for fights in New York City when a group called Children Before Dogs tried to eliminate street and sidewalk pollution in 1971. There were in the city an estimated 500,000 dogs producing an estimated 50,000 tons of dung and an estimated 5,000,000 gallons of urine per year. *Newsweek*, April 12, 1971, p. 95.

11. *Ibid.*, Table 979, p. 620.

12. E. Paul Taiganides, "The Problem of Farm Animal Waste Disposal," in a pamphlet issued by the American Society of Agricultural Engineers to announce the publication of *Management of Farm Wastes, op.cit.* and "The Animal Waste Problem," Nyle C.

Brady (ed.), *Agriculture and the Quality of Our Environment,* (American Association for the advancement of Science, Washington, D.C., 1967), pp. 385–94. Scientists and engineers usually do not deal in such crude comparisons as cubic yards. Human sewage contains much more moisture than animal waste. Animal manures vary from ruminants to swine to poultry because of differences in the animals' digestive systems and types of feed. Such variations mean that more precise analysis deals with such factors as the amount of oxygen required to decompose manure (biochemical oxygen demand). See, for example, S.A. Witzel, E. McCoy, L.B. Polkowski, O.J. Attoe, and M.S. Nichols, "Physical, Chemical and Bacteriological Properties of Farm Wastes (Bovine Animals)," in *Mangement of Farm Animal Wastes, op. cit.* We think the crude comparison by cubic yards is accurate enough to make the point that animal manure is ten times greater than human manure in the United States.

13. Samuel A. Hart, "Processing Agricultural Wastes," paper presented at the National Solid Waste Research Conference sponsored by the American Public Works Association and the U.S. Public Health Service, Chicago, Dec. 2–4, 1963.

14. *Wisconsin State Journal,* March 24, Oct. 2, 1970. As one sign of how fast awareness of the environment developed, the state's Resource Development Board three years earlier while considering a model ordinance to protect shorelines raised the question of pollution of waters by farm animals and was told that "the subject was touchy from both a political and a legal standpoint." *Ibid.,* Oct. 3, 1967. By 1971 the staff of the Dane County Agent included an Environmental Quality Agent.

15. Grant Cottam, "Quality in Biotic Communities," mimeographed, in the record of the Wisconsin Seminar on Environment, Institute for Environmental Studies, University of Wisconsin, Madison, Wis., 53706. All direct quotation will be from this paper and from a supplemental section of recommendations by Mr. Cottam, a member of the seminar.

16. John T. Curtis, *The Vegetation of Wisconsin: An Ordination of Plant Communities* (University of Wisconsin Press, Madison, 1959).

17. F.R. Bowerman, "Introduction," Richard C. Corey (ed.), *Principles and Practices of Incineration,* (Wiley-Interscience, New York, 1969), pp. 5–7. The figures for quantity were cited from Leonard S. Wegman, "Planning a New Incinerator," *Proceedings 1964 National Incinerator Conference,* American Society of Mechanical Engineers; those for categories of waste from American Public Works Association, Public Administration Service, *Refuse Collection Practice,* Third Edition, 1966.

18. Richard B. Engdahl, *Solid Waste Processing, A State-of-the-Art Report on Unit Operations and Processes,* Public Health Service, Bureau of Solid Waste Management, Washington, D.C., 1969).

19. Taiganides, *op. cit., Agriculture and the Quality of Our Environment,* p. 385.

chapter

— 3 —

THE AIR

»»»»»»«««««

METEOROLOGISTS LIVE in difficulty. They must deal scientifically with climate and weather. Other men deal with these subjects intuitively. We know, intuitively, that it is colder than the thermometer indicates when a sharp wind is blowing. Meteorologists must know and explain why we feel colder. They also work in a field in which the weather is described and forecast many times a day in print and by electronics to all who want to know. But weather forecasts are only a small part of the subject for which the meteorologists should not be held responsible.

Meteorologists, to add to their difficulty, are few in number, although the research that needs to be done grows steadily larger and more urgent in social terms. Some 400 definable U.S. meteorologists, with doctorates, must try to learn the facts about the atmosphere which surrounds the earth and what can be done about the weather.

Meteorology up to the time of Newton [writes Reid A. Bryson of the University of Wisconsin Department of Meteorology] was largely descriptive. Since that time it has gradually become largely concerned with the application of "classical" physics (the Newtonian Laws of Motion and thermodynamics) but to a gaseous system where complete observation has not been possible and where the controlled variable type of experiment is out of the question except on a very small scale. The task of observing the atmosphere is immense, for it changes so rapidly that total *synoptic* coverage must be obtained in contrast to the small region-at-a-time *sequential* observations possible on the static lithosphere (the realm of geology), or the one-shot, deliberately prepared experiments of physics and chemistry. This requires international cooperation and an expensive, extensive corps of observer-technicians before even one scientist may study the circulation of the atmosphere. . . . The atmosphere is an enormously complex, rapidly changing vast medium interacting with the sea, the solid earth, and the solar atmosphere.[1]

This hesitant introduction is to say that whatever we say now about the atmosphere as environment may be changed greatly as research and development in meteorology expand, as they will from necessity. Already the science has reached into space to study the radiation of heat from the earth and from other planets and to study from earth's satellites the formation of weather on the earth. Already the jet streams have been mapped, "great atmospheric rivers . . . part of the circulation pattern of the atmosphere that hems the highly productive lands of the world between the subtropical deserts and the frigid subpolar climates and brings the alternating seasons, rain and sun, that are essential to a high level of agricultural productivity and human energy." [2] Already enough has been learned about clouds and storms to allow most meteorologists to assume that some day man can modify his weather. Already research has shown that all the dust kicked up by man and domestic animals, all the smoke and topsoil, all the gases from earth, all the jet vapor trails, will change the climate on earth if the amount continues to increase at the present rate.

The same kind of knowledge is accumulating about microclimate, as meteorologists designate the climate that is immediate to life on the earth. The atmosphere at high altitudes and outer space affect the microclimate—the end would come if clouds, perhaps of our own making, reduced too much our access to the sun—but the climate right at hand is the one that man feels as his environment. It is with this microclimate that we are concerned.

TRENDS IN CLIMATE

The one thing certain about climate is that it is always changing. It changes slowly, and for years because the change was so slow meteorologists searched for the average climate. Then they began to collect evidence of change for recent time—the change for prehistoric time was known—and the emphasis changed from talk of average to talk of change. The evidence comes in tree rings, the first source used for North America, from carbon dating of man's artifacts from archaeology, and from written records. Biological and social phenomena reflect the climate of their time.[3]

Only the highlights of major change need to be mentioned to make the point that climate is always changing. When the last ice disappeared from Scandinavia, England, and North America about 10,000 years ago, there was less moisture in the atmosphere and hence less rainfall. Temperature went up. The dry Sahara began to drive out men and such animals as elephants and giraffes, although a few elephants survived in isolation in Algeria until Roman times, and H.H. Lamb makes the flat statement that Hannibal used Algerian elephants in his Roman campaigns.[4]

Then came a rainy, cool period, around 500 B.C. It was good for the Greeks but bad for the Lake Dwellers of Europe, whose bogs were inundated by rising waters. From about 800 B.C. until Roman times, the Greeks enjoyed a benevolent climate. Not so the Northern Europeans. One belief is that bad climate and storms in the North Seas set the Celtic and Teutonic people to moving from Western and Northern Europe during the first century B.C.

Next the climate gradually became drier and reached a new warm and dry period between 400 and 1200 A.D. This was the time of the Viking voyages and the settlement of Iceland and Greenland, with graves remaining in Greenland deep in ground that is now frozen the year round. There were vineyards in England as far north as York. But while Northern and Western Europe enjoyed the warm, dry climate, the Mediterranean was subject to cold spells that brought frost, and there are eyewitness records of the Nile frozen at Cairo and the Tiber frozen at Rome.

The cold began to return to Europe in 1200 A.D., to continue for two hundred years. Vineyards began to disappear, not only in

England, but in other northern lands of Europe and in higher altitudes. The European colony in Greenland died. Iceland suffered disasters from ice as well as from volcanoes. From 1400 to 1550 Europe experienced a partial return of warmer climate.

Then came three hundred years, 1550–1850, of cold and ice. Glaciers advanced farther than at any time since the Ice Age. Houses were built for maximum protection against winds from all directions. Farms and farmlands were covered by ice in Iceland and from Norway to the Alps. Forests died in Northwest Scotland, an event for which some Scots still blame the English. Until 1831 the bridges of London were built in a manner that restricted the flow of water into and out of the upper Thames and thus reduced the tidal current. The upper river froze, not often in any century but four to eight times a century in the period 1500 through 1700, compared to but once or twice in any century from 900 through 1400.

The years since 1830 have brought a return of the warming trend. Records, now based on measurement by instruments, indicate significant changes. Mr. Lamb concludes, "These may be summed up by saying that the nearly world-wide climatic amelioration—seen at its most rapid in the warming of the Arctic and ice recession—from the 1830's to the 1930's has brought us into yet another climatic phase, partly resembling the warmer periods in the Middle Ages." [5]

The history of climate in Central North America has been pieced together from archaeological records, particularly from those which show biotic remnants, including pollens. This work has been done primarily by three of our own colleagues at the University of Wisconsin: David A. Baerreis of Anthropology and Reid A. Bryson and Wayne M. Wendland of Meteorology.[6] Only the most recent period can be studied from written records and instruments, because the American Indian had no alphabet and the Europeans reached the Midwest in numbers only in the 1800's and later.

It seems clear that the climate changed abruptly at the end of the Late Glacial period. Boreal (northern) forests, which had grown as the glacier retreated, disappeared from large areas of the Dakotas, Nebraska, Iowa, Illinois, Wisconsin, Minnesota, and eastward. Grass and mixed woodlands replaced the forests. By some 8,000 years ago the pattern of mixed woodlands and grass through what are now the Lake States, topped by the boreal forest in what is now Canada, had been established. All the ice sheet was north of the 50th parallel except for a southward dip almost due north of where Cleveland is now.

By 5,000 to 3,500 years ago the northern edge of the boreal forest crept farther north as the ice retreated. About 2,500 years ago the forest became much wetter and the growth of blanket peat, or "upland muskeg," began. Indeed the post-glacial period of climate, very different from the present, seems to have continued in Central North America until about 500 b.c. The climate has been changing on a smaller scale ever since. It was more severe until about 400 A.D.; milder again until about 1300 A.D.; more severe, especially in the period from about 1600 A.D. until the late nineteenth century; and has been getting milder since the late nineteenth century.

CAN WE LOOK AHEAD?

It is easy to say that the climate of North America, and all other parts of the world, will probably not be the same a thousand years from now. It is more difficult to say what it will be and how rapidly it will change. The tempo of change may well be speeded up if man's increased numbers and increased technological activity on the earth continue to grow unchecked and if man's practice of using the atmosphere as a dumping place for waste is not changed. For the first time in the record, man seems to be the chief contributor to the conditions that cause climatic change. Before now the causes have been due to such uncontrollable events as the melting of ice or clouds of volcanic dust.

The fundamental source of climate is the physical fact that heat that comes to the earth's surface from the sun (and to a much lesser degree from the earth's interior) "must be disposed of or the earth's surface will get hotter and hotter." [7] The heat that comes to the earth's surface leaves it through infrared radiation from the earth to space.

Any change in either the receipt of heat or the release of heat will change the temperature of the earth's surface. Thus some meteorologists stress fluctuation in the sun's emission as a cause of change in climate; others stress the earth's reflectivity; and there is good evidence that all are correct. Climatic change is caused by change in the sun's output and also by such conditions of reflectivity as the amount of snow and ice on the earth, the amount of carbon dioxide in the atmosphere, dust, or the kind of groundcover on earth. "Any change which makes the earth a brighter planet, that is, which results in more of the sunlight being reflected, lowers the mean temperature of the earth. Increased cloudiness, snow cover, or dustiness of the atmosphere make the earth a brighter planet." [8]

Fortunately, the transfer of heat from sun to earth and back to space does not occur in a simple "bang to bang" reaction. The process of radiation involves heat that warms the earth by conduction, and a benevolent womb of the atmosphere that re-radiates some of the heat downward; the meteorologists must be concerned with the effective emissivity of the earth as well as with reflectivity. A change in either can change the mean temperature.

Man, the first creature to begin to understand climate, is also the first creature who can change it. He cannot do anything about the fluctuations of the sun, but he can change the earth's reflectivity and its effective emissivity.

He can lay pavement, melt ice packs, and build lakes. He can burn more and more brush and more and more fossil fuels to send more carbon dioxide and smoke particles into the atmosphere. He can allow his goats to overgraze and he can plow up grasslands, both actions that start the soil erosion that sends clouds of dust into the atmosphere. He can put more and more jet airplanes into military and commercial use at high altitudes until their vapor trails form considerable clouds. When he adds his own dust to the dust sent up by volcanoes, the amount of change in the atmosphere is enough to change the temperature on earth.

One recent puzzling fact suggests that man's dust may be more significant than carbon dioxide in affecting the earth's temperature. Carbon dioxide in the atmosphere has increased by some 11 percent since 1870. This makes it harder for heat to radiate from the earth and therefore should cause a rise in mean temperature. The expected happened. From the 1880's until the 1940's the earth's mean temperature rose by 0.7° Fahrenheit. Then, while the amount of carbon dioxide continued to rise, the temperature began to cool in 1940 and by 1960 had cooled by almost a third of the gain during the time of rise. Dust scatters incoming sunlight and reduces the amount of heat that reaches the earth. When a cooling trend coincides with an increase in dust, the dust is suspected as a cause.

Some new evidence indicates that the carbon dioxide may not come from burning fossil fuels, as first suspected, but from vegetation, the slow oxidation of peat bogs, the burning of farm fields, or even from the soil itself. In any case, the warming effect of carbon dioxide will not offset the cooling effect of dust. And man has more control over dust than he has over carbon dioxide.

We can look far enough ahead to say that man should reduce the amount of dust he puts into the atmosphere. Otherwise he will cool the earth, perhaps more than he would wish or find comfortable.

Whatever changes occur from natural or human causes, man on the ground will still be concerned with the basics of temperature and precipitation and with the winds that affect them.

LOCAL CLIMATE

If it seems ridiculous to shift to our climate from the cosmic climate of earth in relation to space and sun, we justify it by the obvious: we do not live in the earth but in Dane and Iowa Counties, Wisconsin. Other men live in similar small areas. The winds we feel are the winds of home, and the climate that makes life easy or hard is the climate just outside our doors.

Most of the people who have lived in these counties all their lives may not think it true, but according to all comparisons with the really severe parts of the world, these counties have an agreeably temperate climate. True, the comparisons are of mean temperatures and amounts of precipitation and avoid the concerns of a man who has to work in a cold high wind or the man who hates to shovel snow. This may well be the point at which emotion and science can never meet, and we look upon fact as irrelevant to feeling.

Our native suspicion of the weather may come from its variability. C.E.P. Brooks, of mild Great Britain, thought it was good for us:

> In the absence of a permanent winter anticyclone the interior [of the northern United States] does not suffer from the intense steady cold of Siberia. Instead, the whole eastern and central parts of North America have a regime of violent alternations of warmth and cold, including the characteristic "cold waves" of winter and "heat waves" of summer. . . . The highly variable climate of most of North America . . . is very stimulating, and probably accounts for the restless energy of the people in the northern part of the continent. . . . Against this must be set the magnitude of the climatic catastrophes—blizzards, icestorms, tornadoes, hurricanes, great floods, droughts and duststorms. These climatic "accidents" receive a great deal of notoriety, and the following pages [in which he recites some accidents] may give the impression that the inhabitants of North America are always liable to be frozen, melted, blown away, washed away, dried up or choked by dust. That would be a gross exaggeration; North America is a large continent and climatic "accidents" are local; probably the majority of the inhabitants go from the cradle to the grave without suffering from any great disaster.[9]

The inhabitants of Dane and Iowa Counties, in their restless energy, demur. They have lived through blizzards and ice storms almost yearly, and tornadoes have struck once in a while. They face cold waves and heat waves every year.

John L. Lambert, a member of the Wisconsin Seminar, gives the basic facts about climate in Dane and Iowa Counties.[10]

"*Wind.* The annual wind rose [a graph that meteorologists use] showed predominantly westerly winds. . . . The lower wind speeds showed an easterly component fairly often, but the higher wind speeds (above 19 MPH) showed almost a prevailing westerly component. During the winter the winds are predominantly from the west northwest, while we are often under the influence of Canadian polar air; during the summer they are from the southwest, while we are often under the influence of maritime tropical air from the Gulf of Mexico; and in the spring and fall our winds are more variable, when the mean position of the polar front lies in our general area.

"Occasionally the area is subject to damaging winds associated with cold-air outbreaks from the northwest . . .

"*Temperature.* The area displays a typically continental daily temperature variation. In the summer (July) the average maximum temperature is about 83° F and the average minimum is about 61° F, giving a daily range of about 22° F. During the winter (January) the average maximum and minimum temperatures are about 27° and 10° respectively, the range being about 17° F. . . .

"Since the area lies in a major storm track, the temperature also fluctuates aperiodically as different air masses dominate our weather. At any time of the year we may be influenced alternately by air from Canadian, Tropical, or Pacific source regions. The relative periods of influence by these three air masses vary throughout the year, thus sometimes providing an extreme diversity of temperature."

[The statement of average temperature does not prepare a startled newcomer for a row of days in which he listens to the forecast, "High expected for today, zero to five below zero."]

"*Precipitation.* One might think that precipitation would occur more frequently during certain hours of the day, especially during the summer, with afternoon and evening thunderstorm activity. Such does not appear to be the case. The occurrence of precipitation is pretty well distributed throughout the day, especially during the summer. . . .

"As is true for most of the northeastern United States, our area receives more precipitation in the summer months than during the winter. For example, the average July precipitation is about 4.5 inches, whereas we receive only a little more than 1 inch during January. The explanation of this variation is found in the air mass frequency for our area. The cold air which dominates our weather during the winter is quite dry . . . but the warm air we receive from the Gulf of Mexico in the summer is heavily moisture-laden. . . . To most people this is uncomfortable because moisture loss (sweating) is essential for comfort at high temperatures, and the presence of large quantities of moisture in the air greatly inhibits this moisture loss from the skin surface.

"The wind speed is a very important factor in the heat loss from human skin surfaces at both high and low temperatures. Unfortunately, in our area, as in many others, the wind speed acts in such a manner as to provide maximum discomfort. In the winter when we want to conserve body heat, we have a maximum frequency of occurrence of high speed winds, and in the summer when we want to cool ourselves, there is a minimum of such higher winds."

This is a climate with about the same benefits and disadvantages as in most of the United States. It gets too hot and too cold for comfort but not for very long periods. It has plenty of precipitation but not too much. It offers variety and thus adds interest to life and a ready topic for conversation.

MODIFICATION

From a combination of stoicism, ignorance, and stupidity man did a poor job everywhere in the cities of the United States in terms of modifying his microclimate. He crowded homes, office buildings, schools, factories, and automobiles onto a small surface. These produce heat and also change the reflectivity of the earth wherever a city accumulates. Then he invented cooling systems and began to modify the climate within a single structure while leaving the city to continue to heat the air above it. It is too late to undo cities already built but not too late to start paying attention to their microclimates when they are rebuilt, as so many are being rebuilt all the time, or when they expand to cover new land surfaces.

In Madison—and other city dwellers can define their problems just by stepping outdoors—our objective in modifying the microclimate of the city is to lower the air temperature in the summer and raise it in the winter. "In the summertime we want to maximize the light reflected toward it, and we can do this with coatings such as paints with specific spectral characteristics. For example, we might coat the roofs of all buildings so that they would reflect a large portion of the solar radiation, but we would leave the sides of the buildings in a less reflective state so as to

minimize the glare on the ground surface below. In the winter-
time when we want to maximize energy absorption, the sun is
low in the sky and most of the radiation falls on the sides of the
buildings and is absorbed." [11] New roofs in southern Florida are
now being painted white.

The presence or absence of moisture for evaporation makes a
big difference in heating or cooling. While the countryside re-
tains moisture in the soil and vegetation, the paved city funnels
its rainfall into storm sewers as fast as possible and only a small
part remains to evaporate. The suggestion is clear. Preserve as
many areas of vegetation in the city as possible, on streets, in
parks, and in residential lots. Scatter trees throughout the city.

More moisture can be provided for evaporation by the old device of allowing water to stand on a flat roof. It has to be kept
clean, and free of mosquitoes, and in cold country it has to be
drained off for winter, but it cuts down the heat.

Finally, for rebuilding and for future building much more attention should be paid to wind. Madison may be fortunate in its
winds. "Wind greatly increases the evaporation from both free
water surfaces and vegetation because the air motion carries the
evaporated moisture away and disperses it. Wind is also an important factor in sensible heat loss from buildings and the
ground surface. A conflict arises here, though, because a permanent structure which shelters an area from wind in the winter
(and prevents sensible heat losses) also tends to inhibit air motion in the summer when the heat losses are desired. The situation may be alleviated somewhat by the fact that the winds in
Madison generally come from the northwest in the winter and
from the southwest in the summer. Thus buildings, city blocks, or
even streets could be arranged in such a manner as to hinder
wind flow in the winter and allow it in the summer."

THE QUALITY OF AIR

Man's social habits have lagged behind technological change.
Man still fights wars, as his biblical fathers did, but with weapons that are the product of the industrial explosion. He treats
the air upon which he depends for life as if it were an unlimited
void to be filled by wastes. His governments and his private in-

terests alike do nothing about pollution until they are forced to by danger.

Nothing illustrates the accumulated threat of such separation of the growth of technology from the growth of social practice as the deterioration of the quality of the air. If a comparison can be made, we would say that in this last third of the twentieth century air pollution is a greater threat to the life of mankind than nuclear war.

The story is an old one. Man began to pollute the air when he began to burn fields, heat houses, operate factories, drive combustion engines, and burn solid wastes. All such activity grew as population grew, and the pollution became more concentrated over cities as the population and its activity concentrated in cities.

At the worst we all die from oxygen starvation. The oxygen in our atmosphere is not a permanent feature, but is rather the result of a delicate balance between its formation in the photosynthetic process and its removal by animals, by combustion, and principally by the slow oxidation of minerals on the earth's surface. This balance completely renews the oxygen supply in the atmosphere in the course of a very small number of millennia. A change of a few percent in the rate of either production or removal could well result in an enormous change in the steady oxygen level.

Although it is unlikely that man or his machines could "breathe" enough oxygen to seriously contribute to upsetting this balance, he could, either through nuclear war or through sheer numbers, destroy enough green plants to seriously decrease the renewal rate of atmospheric oxygen. At present there are no data on which to base a quantitative statement; however, the possibility seems very real. . . .

These considerations suggest that the time has come for man to stop regarding the atmosphere as unlimited and to undertake its conservation.[12]

At best we could achieve true air conservation to insure the balance of oxygen and to keep the air clean enough so that breathing it does not damage health or cause discomfort. The air gets polluted enough from natural causes such as vegetation and wind dust and from the necessary activities of man. It is now essential that man's activities be conducted in ways that will contribute the least to air pollution.

Truly the air is filled with danger. Threats and damage come from burning oil and coal, which we do in prodigious quantities in industry and home heating. They come from burning gasoline —and from not burning all of it—in automobiles, which have become our system of mass transportation. They come from the evaporation of gasoline from the tanks and carburetors of those same automobiles. They come from burning the preposterous tons of solid waste that we produce from our paper-rich and plastic-sanitized culture. They come from industrial processes, such as paper pulp production. They come from dust in the streets and from sprayed fields. They come from exploding nuclear devices in the atmosphere—and to add to our own practices, from the slash-and-burn agriculture of increasing numbers of more primitive people.

The threats and damages can be gases or particles. Winds, terrain, and man-made structures guide the horizontal flow of pollutants. The vertical levels of temperature decide how slowly the gases will rise and disperse, and a city is in trouble when warm air lies on top of cool air so that all the gases and particles are held close to the ground. Nature will dilute, disperse, destroy, or convert many pollutants—if this were not true we might not be alive—but a change in the natural condition that lasts for a day or more means clear and present danger to life and property.

No one knows how many deaths are due to air pollution! There is no such disease. Physicians do not write death certificates that cite air pollution as the cause of death. They cite emphysema, chronic bronchitis, pneumonia, heart failure; and the research on the connection between these causes of death and air pollution is still too tentative to be reliable. It took years of gathering data before the firm connection between cigarettes and lung and heart troubles was accepted. It will take years to determine the connection between air pollution and lung and heart disease.

The study will be more difficult too. The cigarette smoker carries his environment with him. The non-smoker changes his air environment several times a day and sometimes for years at a time. The few tentative studies suggest that a connection between air pollution and lung cancer does indeed exist. The incidence of lung cancer is higher in urban than in rural areas, higher among British immigrants to South Africa than among native born whites, although the native born whites smoke as much or more than the British immigrants. Other studies have shown that air pollutants can induce cancer in laboratory animals.[13] Then there were the disasters in the Donora Valley of Pennsylvania, in London, and in the Meuse Valley. The abrupt rise in the number of deaths with the increase in pollution left no doubt that the deaths were caused by foul air. There is still some doubt about which pollutants were most responsible, but no doubt about the cause of deaths.

A listing of known pollutants is the best way to make the point that the air we breathe is contaminated and if the portion of pollutants (expressed as parts per million, ppm) rises, damage follows.[14]

Sulfur and its compounds. The most common form of sulfur in the atmosphere is sulfur dioxide. It comes from burning coal, oil, and solid wastes, from sulfuric acid plants, and from metallurgical processes that involve ores that contain sulfur. Some of the sulfur dioxide is oxidized further to become sulfur trioxide. When sulfur trioxide combines with water vapor, it becomes sulfuric acid. Sulfur dioxide in its usual quantity is not proved to be fatal, but when it combines with aerosols (solid or liquid particles suspended in a gas) or when it changes to sulfuric acid it becomes toxic to a degree that makes it suspect as the cause of death in the three disasters, Donora, London, and Meuse, from which most of our knowledge of mortal effects comes. Sulfur dioxide also damages vegetation, and it and sulfuric acid both cause damage to metals. Sulfur dioxide and sulfur trioxide also damage limestone, marble, slate, and mortar, all components of the world's treasured buildings and sculpture.

Mercaptans (any of a group of compounds all of which contain a common ingredient in the same way the alcohols form a series) and hydrogen sulfide are the other sulfur compounds produced copiously by an industrial society. They come from oil refineries, coke ovens, paper pulp mills that use the Kraft process, tar distilleries, natural gas refineries, and from plants manufacturing viscose rayon. They also come from dumps and sewage plants where treatment is inadequate. These are the stinking compounds, and they stink at very low volume, so low that odor nuisance can be caused when the concentration is 10 to 100 times smaller than the smallest concentration of sulfur dioxide that is detectable by smell. People can smell hydrogen sulfide in concentrations of about 0.035 to 0.10 ppm.[15] Silver and copper tarnish from hydrogen sulfide and house paint that contains lead will turn dark from it.

Carbon monoxide. Each automobile in Los Angeles County pollutes the air with an average of more than five pounds of carbon monoxide each day. Cars in other cities produce the same amount but in more concentrated locations. People feel dizzy, listless, and get headaches at a concentration of 100 ppm. Near Oxford Circus in London the concentration is frequently more than 100 ppm. In New York the U.S. Public Health Service finds that within 100 feet of heavy traffic the concentration of carbon monoxide is three times the amount that began to slow the mental and physical reactions of Cincinnati firemen who volunteered to take T-crossing tests while breathing air that had controlled amounts of carbon monoxide fed into it.[16] There is still a great deal to be learned about the amount of carbon monoxide in the air and about its effect upon people, short of death, at a really high concentration.

Carbon dioxide. This gas is listed among pollutants only because man's burning of fossil fuels increases the amount in the atmosphere to such an extent that the normal balance is upset, with possible results that cannot be foreseen. The effects on climate, as we saw earlier in this chapter, are not yet clear because the warming of "the green house" may be offset by the cooling effects of man's dust. Carbon dioxide is emitted naturally by plants, volcanoes, and by man and other animals. Of more concern is the amount of additional carbon dioxide that man releases into the air by burning coal and oil.

Oxides of nitrogen. The combustion of coal, oil, gas, or gasoline at high temperature fixes atmospheric nitrogen to produce first nitric oxide and, from a reaction with oxygen, nitrogen dioxide. The conversion of nitric oxide to nitrogen dioxide is accelerated in sunlight, especially when organic material is present, as in smog. Since burning at high temperatures is typical of an industrial-internal combustion technology, the oxides of nitrogen are serious pollutants in many communities. They have the usual effect of threatening health and damaging vegetation. When nitrogen dioxide reacts with water vapor or raindrops to produce nitric acid, metal surfaces can corrode.

Photochemical pollutants. The hazy condition known as smog is formed when oxides of nitrogen or sulfur dioxide mix with organics in the presence of sunlight. This is the most familiar form of air pollution because it can be seen. It damages plants and irritates the eyes; it causes rubber to crack and it reduces visibility. The most common organic gases that are present in sunlight to join with the oxides of nitrogen and sulfur dioxide are hydrocarbons, aromatics, and aldehydes. These arise from the incomplete combustion of all kinds of fuel, including rubbish and brush, but automobile exhaust emissions are by far the largest source. The organic gases also come from refining and other gasoline processing and from factories that house driers, ovens, and furnaces.

Particulate matter. Both liquid and solid particles add to the gaseous burden of the air. The large particles settle to the surface sooner than the small ones. Ash, dirt, and soot, are obvious examples. Less obvious are the metallic elements, the most frequent being silicon, calcium, aluminum, iron, magnesium, lead, copper, zinc, sodium, and manganese. Particulate matter comes from virtually all the activities of modern man, including his prodigious use of automobiles, and from such natural sources as winds or volcanoes. The greatest concern for particulate matter in air pollution, however, involves the matter sent up by man. Particulates cause about the same damage as the gases. They are threats to health. They are nuisances. They lower visibility. They injure vegetation. The most ominous as threats to health are lead and other deleterious metals. The Air Conservation Commission devoted an entire chapter to these metals.[17] Tetraethyl lead added to gasoline passes into the atmosphere from auto exhausts and from evaporation, most of it as lead bromides or chlorides. Lead particles in the street dirt of New York increased from 1,190 ppm in 1924, the year after tetraethyl lead was first introduced as a gasoline additive, to 2,650 in 1953.[18] A great deal still has to be learned about the physiological effects of metals in the air.

Fluorides. Most of the fluorides in the air come from such raw materials as ores, coal, clays, and soils, although some come as well from certain industrial processes, such as those used in the aluminum and fertilizer industries. They are harmful pollutants because they cause damage to certain plants and to animals that eat contaminated plants. Cattle lose weight, have trouble with teeth and bones, and give less milk.

Economic poisons. This category includes all those man-used materials that are toxic but deemed so useful that they are used in spite of the risk. They kill pests, insects, fungi, weeds, rodents, spiders, and worms. They are supplied from the air and on the ground, to forests and farm fields, in homes and small gardens, in buildings and parks. Remarkably little is known about how many of the economic poisons constitute how much of a hazard in air pollution.

Radioactive pollution. So much attention was given to fallout during the period before the Test Ban Treaty was signed that little attention was directed to other sources of radioactive pollution. And for that matter, China and France did not sign the treaty, and fallout is still a considerable source. The other sources are the reactor-fuel cycle, the use of nuclear energy for propulsion, and the use of radioisotopes in industry, medicine, agriculture, and scientific research. To these sources many would add another: the nuclear accident that possibly can occur and spread radioactivity over populated centers and surrounding lands. Some cautious observers think that such an accident is bound to happen sooner or later and place little faith in the statistics of probability when they consider the awesome and improbable effects of floods and tornadoes and the imperfectability of technical man.

Short of an accident, the normal operation of a reactor produces fuel elements and coolants that are possible sources of air pollution. Radioactive gases are released into the air, more from air-cooled reactors than from water-cooled reactors. Chemical plants that reprocess atomic fuel, separating the unfissioned uranium and plutonium from radioactive waste, also emit gases. As the number of reactors for the production of electrical power increases, this source of radioactive pollution may become much more significant.

Nuclear propulsion of submarines is already here. Propulsion reactors for space are in the research and development stage. All propulsion reactors can contaminate, just as any stationary reactor can. (In addition, isotopes used for supplying energy on space vehicles can escape into the atmosphere when a vehicle either fails to go into orbit or whenever an isotope does not burn up completely on re-entry into the atmosphere.)

Not much radioactive pollution gets into the atmosphere from any one hospital or research laboratory. The total from a large number of such institutions, however, could become considerable.

SOLUTION

These are the chief pollutants. They are all harmful to man and other forms of life. They are all the results of man's activity in a complex technological society.

What to do about them, as always, will involve both technical and social solutions, and, as usual, the technical solutions will be found easier to accomplish than the social solutions. Considerable work is already underway on the technical side, more work indeed than can be summarized here. It includes smoke stack control; re-use of gases; conversion; extra high temperature combustion and the reduction of solid waste to crystals; compression of solid waste for landfill instead of burning; controls over automobile exhausts; efforts to develop steam, electric, or fuel cell engines to drive automobiles; and other methods that reduce the amount of contaminants in the air.

Whenever life becomes so unpleasant or so dangerous that a considerable number of people complain, city councils have responded with legislation aimed at warning and control. Congress has adopted a law to require controls on automobile exhausts. The striking thing about a visit to any of the cities with high pollution is, however, the remarkable patience of people who are accustomed to an industrial-automotive society. The residents of

New York, Los Angeles, or Chicago, to name only three, will endure physical irritation, stenches, and pulmonary distress in preposterous degree without complaining and, apparently, without thinking too much about it. More of them each year will die from the results of air pollution, but because their ailment is not so diagnosed they will not know what really killed them.

GUIDELINES FOR CONTROL OF AIR POLLUTION

Since it is unfeasible to go into detail about the technical methods of control over air pollution or about the laws that have been adopted in the nation and in various cities, we will instead suggest certain premises that should guide individuals, institutions, even society itself in the fight to save our usable air.

1. The goal of control should be to maintain the quality of life as we know it. Certainly it would be possible to place each human at birth in a man-made environment and keep him there until death. We have done this for high-altitude and space flights and for underwater living. We could perhaps even take off the helmets and suits when we entered buildings that were themselves air conditioned capsules. This is not the kind of life that we want, however. We want a mother to continue to talk to her baby face to face and not by radio. We want to see faces on the street. We want the earth's atmosphere to be used by all living things. The beasts of field and forest are valuable neighbors, and so are the pre-industrial people. Stone, marble, and metal works of art and architecture are valuable things to save. To all statements that technology will find a way to save man from himself, the answer should be: We don't want it; we want to breathe the air that nature gave us, in comfort and in good health.

2. All living creatures are entitled to live as long as possible. In the case of man, who can do something about his environment, the shortening of life because of polluted air is morally indefensible. It is nothing less than social murder. Each case of chronic bronchitis, each case of emphysema, each case of other pulmonary or cardiac disease that is attributable to the contaminated air—and not consciously risked by smokers—is an indictment of the whole society and all its institutions, both public and private.

3. Air is the property of everyone. It does not belong to private industry, to automobile owners, to farmers who spray crops, or to any single person or group. It is owned as much by the primitives of New Guinea as by executives in New York. All other property rights are subordinate to the property right of the whole human race in the atmosphere of earth.

4. The cost is never too high for life itself. All debate about whether we can afford to clean the air is futile and foolish. If we do not clean the air—and reduce its contamination—we die in larger numbers until we are gone. Only the silliest of economic reasoning would argue that we cannot afford to spend the $30 or $50 billion dollars spread over ten years that would prevent the present dying and eliminate the air we breathe as a cause of death in the future.

5. The cost will always be spread among most of the adults in the nation. If we spend public funds, we pay for it in taxes. If private firms spend private funds, we pay for it in the prices we pay for the goods or services produced. It seems useless to argue about what kind of money is used to pay for the changes needed. Funding becomes mainly a question of how to do it with maximum speed and lowest carrying charge. If public funds can ensure the quickest and cheapest action, use public funds. If private funds can better get these results, use private funds. Or use a mix of public and private.

6. The goal is always the wisest use of the air, not disuse. Some discussions of air pollution make the point that man himself pollutes the air with carbon dioxide each time he exhales. Vegetation gives off carbon dioxide. Cities and automobiles emit heat. In other words, some kinds of pollution are inevitable in terms of the desirable use of the atmosphere. The danger comes from the harmful pollutants in amounts that exceed the most cautious tolerance for safety. They are the gases, chemicals, and particles catalogued above. They have already entered the air in damaging volume, and they must be reduced before we can begin to define the safe uses of the air.

If both public and private decision-makers, and just plain citizens, will follow these guidelines, technology can save us.

NOTES

1. Reid A. Bryson, "Man and the Earth," Oct. 21, 1964, a mimeo-graphed paper available to the Wisconsin Seminar on Quality of the Environment.

2. *Ibid.*

3. For the period of tree rings and geological analysis, see C.E.P. Brooks, *Climate through the Ages* (McGraw-Hill, New York, 1949), Chap. XXI. For the period after carbon dating, see H.H. Lamb, "Our Changing Climate, Past and Present," in H.H. Lamb (ed.), *The Changing Climate, Selected Papers* (Methuen & Co., London, 1966), pp. 1–20. We have relied on Lamb for what follows.

4. Lamb, *ibid.*, p. 6.

5. *Ibid.*, p. 13.

6. David A. Baerreis and Reid A. Bryson, "Climatic Episodes and the Dating of the Mississippian Cultures," *The Wisconsin Anthropologist* 46(4):203–20, Dec. 1965, and Reid A. Bryson and Wayne M. Wendland, "Tentative Climatic Patterns for Some Late-Glacial and Post-Glacial Episodes in Central North America," in William J. Mayer-Oakes (ed.), *Proceedings of the 1966 Conference on Environmental Studies of the Glacial Lake Agassiz Region,*" Life, Land and Water," pp. 271–98 (University of Manitoba Press, Winnipeg, 1967).

7. Reid A. Bryson, "All Other Factors Being Constant . . . A Reconciliation of Several Theories of Climatic Change," *Weatherwise,* 21(2):56–61, 94, April 1968.

8. *Ibid.*

9. C.E.P. Brooks, *Climate in Everyday Life* (Philosophical Library, New York, 1951), pp. 115–18.

10. In a mimeographed paper, "Climate of Dane and Iowa Counties," in the record of the Seminar, filed with the Institute for Environmental Studies, University of Wisconsin, Madison, Wisconsin 53706.

11. John L. Lambert, "Urban Microclimatic Modification," mimeo-graphed paper in the record of the Seminar, Institute for Environmental Studies, University of Wisconsin, Madison, Wisconsin 53706. Further direct quotation is from this paper.

12. *Air Conservation,* the report of the Air Conservation Commission of the American Association for the Advancement of Science, (published by the Association, Washington, D.C., 1965). This book is the best single source for laymen.

13. *Ibid.,* pp. 142–43, citing the original articles.

14. *Air Conservation, ibid.,* lists the pollutants in its table of contents and devotes a chapter to each one. We have relied upon *Air Conservation* for this capsule summary. Other listings of pollutants differ; for example, the U.S. Public Health Service reports on some twenty-five, including metals, in its annual *Air Quality Data from the National Air Sampling Networks and from Contributing State and Local Networks,* Air Quality Section, Robert A. Taft Sanitary Engineering Center, Cincinnati.

15. *Ibid.,* p. 69.

16. *Ibid.,* pp. 73–76. *Newsweek,* Jan. 23, 1967, p. 87.

17. *Air Conservation,* pp. 124–33.

18. *Ibid.,* p. 125.

chapter

—— 4 ——

THE WATER

»»»»»»««««««

Mᴀɴ ʜᴀꜱ ᴀʟᴡᴀʏꜱ taken water for granted. He had no other choice. He could not live without the stuff. He needed it to provide food. He found pleasure in its gentle rains and terror in its violent moods. Life and death were tied to water. Diversity was here in water.

Today writers who deal with water tend to become rhapsodic when they think of its universality, and few resist the impulse. Water makes tears and the cells of life; furnishes power for machines and for sheer force; water is the rills that flow to oceans; it recalls the origin of life in a dim moist past. The wonder of it en- courages poetry.

As man learned more science, he learned more about water, including one odd fact that pops up more than once. Most water scientists think that no new water is being added to the amount now on earth. For water, the earth is a closed circle. Any water that starts toward the sun is returned to earth's surface and does not escape into space. Water is recycled by nature. Every drop of it is used and re-used. This led some writer with imagination to say that Cleopatra's bath can be in some city main today, and he put into the conversation another item that has to be suppressed at table.

THE NATURE OF WATER

Scientists and engineers agree on most of the main facts about water.[1] Briefly, these facts are:

• Water in all its forms, as a solid in ice and snow, as a liquid, as a gas in steam or water vapor, is all part of the same system.

• Most of the earth's water is in the oceans, an estimated 97.2 percent. The next largest amount is in the Arctic and Antarctic icecaps and in glaciers, some 2.15 percent. Subsurface water accounts for 0.625 percent; surface water (not oceans) for 0.017 percent; and the atmosphere for 0.001 percent.[2]

• The hydrologic cycle keeps all water in circulation. Begin the cycle with precipitation—rain, snow, hail, or sleet. Some of it evaporates as it falls. Some of it stays on the surface of the earth where it runs into ponds, lakes, rivers, reservoirs, and oceans. It evaporates from all these surface waters.

Some of it soaks into the ground, an act called infiltration, and becomes what is known as ground water, which is stored in aquifers, the rock or rock formations that bear water. A whole book could be written on ground water. It makes a most complex system, depending on the structure of soil and rock in any one place. Ground water returns to the atmosphere in these ways. It is taken into vegetation. It penetrates soil. It emerges as springs. It is taken out of the ground by men who dig wells. It seeps into oceans. From all these sources water that has been through the ground route gets back to where it is affected by the sun and re-enters the atmosphere.

It can pass through plants and animals on the way, but it comes out through their pores and transpires instead of evaporating. In the case of animals, including man, it emerges from the body as fluids and stays in the cycle. Evaporation and transpiration gather all the water that ever existed into clouds, from which precipitation starts the cycle again. It is a perfect example of feedback, a wonderful work of nature, and it confirms a poet of an age before science who made Ecclesiastes the Preacher say, "All the rivers run into the sea, yet the sea is not full; unto the place from whence the rivers come, thither they return again." Cleopatra's bath water is in there somewhere.

• Ordinary household use of water is minor compared to other uses. Drinking, cooking, bathing, and washing, consumed 32 billion gallons daily in the United States in 1965. By 1980 it is estimated that this figure will rise to 39 billion gallons daily. Within

the household the flush toilet is the biggest user—41 percent of
the total. In declining rank after flushing are bathing, 37 percent; kitchen uses, 6 percent; drinking, 5 percent; washing
clothes, 4 percent; housework, 3 percent; watering gardens and
lawns 3 percent, and car washing, 1 percent.[3] If man can learn
another way to dispose of body waste and to bathe, he could become a very minor drain on his household water supply. These
two activities account for 78 percent of all household water used.
Even so, man at home is a minor user of water.

Agriculture consumes the most water—148 billion gallons
daily in 1965 and a projected 178 billion gallons daily by 1980.
Irrigation of dry land takes enormous amounts of water. It has
been practised since the early days of America and has been encouraged by the national government since the Bureau of Reclamation was created in 1902. Steam powered electric generating
plants are the next biggest users—119 billion gallons daily in
1965 and an expected 162 billion gallons daily in 1980. The production of electric power is doubling each decade in the United
States. Industry is the next big user, 73 billion gallons daily in
1965 and a projected 115 billion gallons daily in 1980.[4]

Water is not only essential to plants and animals; it is also the
most valuable and most versatile natural resource for an industrial society. It is a solvent, catalyst, and cleanser. Turned
into steam it is a force. It is a means of transportation and a
means of disposal. It cools, dilutes, and disperses other substances. It can be used to distribute heat. Any discussion of water
in the environment must recognize that water is the most essential resource for an industrial society, including agriculture that
produces the abundance of food and fiber for the good life. The
concept of conservation of water means that maximum use with
safety for the future is the aim.

• Whether enough water is available depends upon the answers to several questions. For what purpose? It is one thing to
talk about water for a few people and another thing to think of
water for several million people and a great industrial center.
Southern California had no water problem before millions of
people moved into a region that was never designed by nature
for them. If the climate is cyclical, what stage is it in? In 1915

there was more water farther west in the Great Plains than there was to be in 1934. The Great Drought which coincided with the Great Depression in the 1930's was a result of cyclical climatic conditions, although the poor farmers who settled on the Plains during the time of rain and who had to desert their homes in the time of drought did not know it, nor did many meteorologists, because the history of recent climate change was then a new sub-field. Of what quality is the water? Is it fresh or saline, hard or soft, too polluted to be used economically for human or industrial uses?

Man can overuse the water supply anywhere in the United States where he adopts intensive practices in agriculture or industry or where he piles up population beyond the potential water supply. The potential supply is adequate with proper management for extensive use in the glaciated areas of the Northeast and Midwest and in the Columbia Plateau of the Northwest.

A heavy demand will reduce the available supply in any other part of the nation. In some parts, in the Rocky Mountains, the Northern Great Plains, and a narrow strip along the Pacific Northwest Coast, in the unglaciated central plateaus and low-lands, the supply is just about adequate for the needs of domestic and livestock use. In some regions, the Appalachian and the Piedmont, the basin and range areas, the potential is better, but the demand still has to be kept within limits for safety.[5]

• While the best informed speak of a "famine" or a crisis in supply related to demand, they also talk about, and work on, schemes to find new sources of water. They think that the development of new sources is technically feasible and economically justifiable. The new sources most often discussed are conservation, re-directing the flow of streams, and desalting the sea or inland brackish waters to get fresh water to where it is most needed.

• The recycling of water is the great natural act of conservation, so perfect a system that it is awesome as a pattern for all other parts of the environment. The trouble comes when man damages the quality of the water in the cycle, sometimes so badly that the water becomes unfit for certain uses. Over-pollution, it is agreed by all observers, is serious in all parts of the United States, but it can still be arrested and turned around if we are willing to spend the money and to cease some wasteful practices.

Pollution is not a pretty thought. The *Encyclopaedia Britannica* lists the following pollutants: excreta, organic industrial wastes, infectious agents, plant nutrients (nitrogen and phosphorus), pesticides, waste minerals and chemicals, radioactive substances, heat, fluorides above a proper level, and accidental spills, which may be anything from crude oil on the beaches to deadly chemicals from a highway truck accident.

A well-informed, articulate member of Congress makes another, but similar, list: ordinary sewage and related organic substances, disease carrying infectious agents, chemical nutrients of plants, synthetic-organic chemicals, sediment, radioactive substances, inorganic chemicals and mineral substances, heat.[6]

The truth is that, along with Cleopatra's bath, a glass of water may contain harmless (we hope) amounts of dread and dangerous, not to say nasty, ingredients. If one should go swimming in most lakes and rivers one would bathe in the stuff untreated for safety. But drink the water only after treatment, except in a few remaining lakes and streams not yet over-polluted.

The most common term used is pollution. We used the word over-pollution to introduce this section then reverted to the more common usage. Over-pollution causes the crisis. Conservationists see water as a resource. One of its uses is to dispose of certain kinds of wastes. This is a legitimate and reasonable use of water when properly managed. The raging misuse and mismanagement of water for the disposal of wastes, almost as if there never had been a future but only the present, is customary in the United States. It is taking the nation on a desperate slide into disaster.

Of all these elementary facts about water in the United States, in general the most pertinent and the most urgent in relation to the environment are pollution and the excessive demands for water in certain places, making some note of the possibilities of new sources of water. The rest of this chapter will deal with these subjects.

THE CAUSES OF WATER POLLUTION

All surface water is polluted to some extent. The high mountain brook has collected some minerals from surface runoff. It would perhaps reveal upon analysis the sediment dropped in fallen airborne dust, which may have carried any variety of things, from pesticides to radioactive particles. Excreta of wild animals might be present. Yet a wilderness camper is safe when he drinks this water, and no sane man would say that the brook is over-polluted. The reason for the camper's safety, and for making the distinction again between pollution and over-pollution, is that the mountain brook is performing the normal and natural function of transporting waste.

To a larger stream man can add waste from urban-industrial sources without damage and be safe in downstream use of the water if he does not overload the stream. We are brought back to the management of water as the essential question. Excessive pollution comes from dumping too much waste into streams.

Moving water will cleanse itself if given a chance. It will dilute chemicals. Some of the bacteria in waste will die of natural causes in a short time. Fortunately for man, this is true of the parasites that cause diseases of his digestive tract. As water flows it is recharged with oxygen from the air and from photosynthesis in plants that grow on the surface of a stream or on the bottom.

Oxygen is the great cleanser. It dissolves (oxidizes is the correct word) solids and kills bacteria. How quickly a stream will clean itself depends on its rate of movement, the volume of water, the amount of waste put into it, and the intervals between dumping waste. No stream can clean itself if waste is put into it before it has had time to get rid of the waste injected upstream.

As population and industry grew, as cities became the main way of life, so much waste was dumped into streams that none could cleanse itself before the next urban complex along the shore dumped some more. Added to urban-industrial waste was the runoff from farms, by now covered with fertilizers, pesticides, and livestock excreta.

Citizens at play and at work added their bit too. Their boat motors leaked oil and gasoline; their boat toilets dumped untreated sewage into rivers and lakes. Most of the humans aboard threw cans, bottles, paper, anything disposable, overboard. If the humans lived in a riverside cottage, they emptied sewage directly into the river or, if not directly, they were likely to send it through an inadequate septic tank and a septic field laid in the wrong soil, at the wrong place, then into the river by seepage.

Such uses of the rivers for waste disposal did no harm when Indians and settlers were few. They would do no harm today if cities were small and few. But cities are large, farms are different, and the larger population means that recreational use of the rivers and shores further fouled the water.

Even at best, when one city dumped sewage that could be handled by the river before the next city added more, the river's job was heavy, and the worm's close view of what occurred would have been a shock to laymen. The water scientists knew. One of them, John Bardach, in an excellent book written for laymen, gave a closeup of a river at work disposing of sewage:

> Usually a river will go through several distinct stages as it flows away from a source of organic pollution. Above the source, the stream may be healthy and clear, with a community of insect larvae, snails, crayfish, minnows and other small fishes, and perhaps a few bass. . . .
>
> At the junction with the sewer, where the zone of pollution begins, the water turns bluish and turbid, there are bits of floating sludge, and the air smells of hydrogen sulfide and methane. Next comes a zone where the animal species are limited to those adapted to cope with a severe reduction or total absence of oxygen. Among these are the rattailed maggot, *Eristalis tenax*, and the wriggler of the sewage mosquito, *Culex pipiens*, both of which are equipped with a snorkel-like air tube that pierces the surface film and takes in atmospheric oxygen. Red sludgeworms of the genus *Tubifex* will also be abundant. These worms, of which there may be several thousand per square inch, make a heavy demand on whatever residual oxygen the river still has to offer. In addition, they pass settled organic matter through their guts at a staggering rate: an individual worm, although it is only an inch and a half long, and no thicker than a needle, may accumulate up to six inches of fecal threads in a day.
>
> Water becomes recharged with oxygen while it flows, and attached or suspended aquatic plants contribute to the oxygen supply through photosynthesis. But photosynthesis goes on only in daylight; at night the plants themselves have to take up oxygen to live. It may therefore happen during the night that fish tolerant of low oxygen conditions, such as carp, which come upstream to feed on sludgeworms in a zone where the recovery of oxygen takes

place during the day, die by thousands of suffocation because there is no longer any oxygen at all.

At some distance below a sewage outflow the nutrient-rich environment of partly oxidized compounds produces a third zone, well suited to the growth of such attached algae as *Cladophora*, which form long fronds and color the river bright green. If the flow slows down at that point, the great supply of nitrogen and phosphorus may lead to a condition in which the water resembles a green broth, with the color and consistency of pea soup. . . .

At last, many miles below the town, in the normal course of events, the river returns to a state that may be described as clean. . . .[7]

EUTROPHICATION

Still water is another matter. The chief concern for the pollution of still water in the United States is the destruction of freshwater lakes. Some day perhaps there will be equal concern for ponds and wetlands, but the crisis of lakes now takes nearly all the attention.

If Americans are still here in A.D. 2100, they will be able to read a history of their preposterous misuse of lakes. The glaciers left us well fixed with a profusion of beautiful lakes, including the greatest freshwater seas on earth, the Great Lakes. New York's Finger Lakes were striking in their setting, so were the lakes of Michigan, Minnesota, and Wisconsin. A map of northeastern Minnesota was speckled blue from so many lakes.

By the middle of the twentieth century all the lakes beside which man had built cities were over-polluted, except for a few. Many of them were unsafe for swimming, even parts of the Great Lakes near certain cities. And life guards were posted at bathing beaches at Sheboygan, Wisconsin, on Lake Michigan, not to save swimmers in distress, but to keep people from going into the water.

Lakes receive the same wastes as streams. Depending on the size of the lake and the circulation of the water, lakes can dilute some of the chemicals, as do streams; and some of the bacteria, including those most harmful to man's digestion, will die in a short time. Oxygen will do its work if the lake is not overloaded.

Always some of the dangers will persist, especially from the "hard" detergents and pesticides which take a long time to degrade. The circulation of water in a lake is imperceptible to the eye and can be detected only with sensitive instruments. All movement of contaminants in water obviously is much slower in lakes than in streams.

The word "eutrophic" first meant well nourished, possessed of good nutrition, and was so defined in the *Oxford English Dictionary* through the 1950's. When the plight of lakes was found to be acute during very recent years, the word came to be limited to water and to the bad effects of good nourishment! When the new *American Heritage Dictionary of the English Language* appeared in 1969 it defined the word: "Designating a body of water in which the increase of mineral and organic nutrients has reduced the dissolved oxygen, producing an environment that favors plant over animal life." The noun is "eutrophication," the development by which waters become eutrophic.

Eutrophication is the most striking and devastating thing that is happening to American lakes unless plain chemical and bacterial hazards to health near the big cities can be regarded as equal.[8]

Very simply, all organic sewage is nourishing to plants. Even after treatment in municipal plants the sewage effluent that remains is rich in phosphorus and nitrogen. Surface runoff in storm sewers is rich in phosphorus and nitrogen, carrying fertilizers from lawns, debris from the streets, detergents from car washing, the dung of animal pets, and all the other variety of organic and chemical waste that comes from the urban-industrial way of life or that falls from polluted air. All the surface runoff from farm land carries chemical fertilizers and animal excreta into the streams that feed the lakes or directly into lakes that adjoin farms. Industries discharge more chemicals into lakes, many of them good nourishment for plants.

Eutrophication is a natural condition of lakes. Man speeds up the process when he dumps waste into them, or allows surface runoff to enter them. Phosphates seem now to be the chief reason for faster eutrophication. They are found in human and animal waste and in nearly all detergents, most of which end up in

streams and lakes. There is some hope that a new method of re-
moving phosphates at sewage treatment plants will be found,
leaving only the phosphates in surface runoff as the accelerator.[9]

Scientists still have much to learn about eutrophication. In the
meantime, all of us can see and smell the results. When plant nutrients enter a lake they stimulate the growth of plants on the
bottom and of algae on the surface. These plants require oxygen.
In time there is not enough oxygen for both fish and plants. The
fish begin to die, the game fish going first because trout, pike,
and bass require more oxygen than carp, bullheads, and perch.
Some of the dead fish wash ashore and stink. The algae form
into thick clusters on top the water and make swimming or boating unpleasant. When the algae and other plants die, they too
begin to stink. There is no proved danger to man in eutrophication (only some questions about eye and nose irritation), but an
aged lake is lost as a source of good fish, as an agreeable place to
swim, and as a pleasant thing to live beside.

THE MADISON LAKES

Again a very specific illustration of all that happens to lakes
lies at our door. Dane County was endowed by the glacier with
four lakes in a chain. The largest, Lake Mendota, is surrounded
by Madison and its suburbs and by homes in septic tank exurbia.
Mendota is aging rapidly.

The natural eutrophication of Lake Mendota was noted in
early diaries after the white man settled, although it was not
called by the name, only associated with a bad odor. Nonetheless,
twenty-five years ago the lake was, most of the time, a pleasant
place for clean swimming. Dedicated fishermen, and boys, could
still take pike and other game fish. Once during that time two
boys caught a sturgeon, but this was never explained for there
was no way for a sturgeon to get from its usual waters into Lake
Mendota unless it walked or was carried overland.

Man built the city of Madison and smaller cities in the Lake
Mendota basin. Since Mendota emptied into the Yahara River,
which filled the other three lakes, this meant that the whole lake
system was affected by the growth within the Mendota basin as
well as by development around each of the other lakes.

As usual, man took water for granted, with one exception. The University of Wisconsin, which occupies about three miles of the south shore of Lake Mendota, developed a strong program of research in limnology, the science of lakes, ponds, and streams. The Wisconsin scientists used the lake as a prolific sample and laboratory combined. A local boast was born and grew that Mendota was the most studied lake in the world.

The boast had one grave effect. It lulled scientists and laymen alike into thinking that the answer to nearly any question about the lake would be available when needed. But the research had all been "basic," in the custom of university scientists. Each man had satisfied his own curiosity in his one discipline. The fish biologist had studied fish; the meteorologist had studied winds, waves, and temperatures. The chemist had stayed within water chemistry. There was little cross-communication and no summing up. An amphibious personnel carrier, in its last ungraceful days as army surplus, appeared for a project in water chemistry. It was named "Entropy." The name might have been given to the whole of research on Lake Mendota. It was random. It was uncoordinated. It did not supply the critical answers needed when eutrophication reached its climax. The same would have been true of research at any other good university.

In the late 1960's a Water Resources Center was established by the State and the University of Wisconsin. It quickly adopted a program in eutrophication. If this program moves with coordination, there still may be hope of saving the lake, provided all the politics and economics move with it.

As usual, politics and economics are more the handicap than science and technology insofar as Lake Mendota and other dying lakes are concerned. Science and engineering, given the money and coordination, could locate the sources of phosphates and nitrates and plan ways of keeping them out of Lake Mendota. The barrier would come from the disparate and dissenting units of government involved and from the notions of false economy that would be shared by both public and private bodies when they face the cost of repair and prevention.

Some bold, and some mild, efforts have been made to save Lake Mendota. The State Legislature in Wisconsin was aware of the importance of the environment long before the national press, the Congress, and so many citizens became aware of it. Gaylord Nelson, who was one of the early leaders of the national effort as a United States Senator, had been a leader in conservation as a member of the State Legislature and as Governor of Wisconsin before he was elected Senator.

Whatever is done to save the Madison lakes will more likely be done by the national and state governments than by the city, county, and township governments or by the special districts, those inter-governmental bodies, such quiet, inconspicuous anomalies as water and sewer districts or soil conservation districts.

First, the State Legislature adopted a law that required the City of Madison to build a main sewer line that would bypass the lakes and dump into a creek below them. Then it adopted a law that required certain small towns to stop putting sewage into the lakes. The state courts found a reason to hold the law unconstitutional, but by the time the courts acted, a statewide law on water quality, adopted in a joint federal and state attack on water pollution, had given the State much stronger authority over both private and public offenders.

In early 1970 the national Agricultural Stabilization and Conservation Service announced that Dane County would be the locale for a pilot project in the abatement of farm pollution of Lake Mendota. The agency, a division of the United States Department of Agriculture, administers price support and soil practice programs on the farms of America. It reaches every farm and ranch in every county of the nation and could exercise considerable leverage in changing farming practices. It pays benefits according to land usage as well as for control of production. In the Dane County trial, it will pay up to 80 percent of the cost of anti-pollution practices. At the beginning, given the lack of technology, the stress would be on the construction of manure pits and other practices which permit manure to be used as fertilizer and thus prevent its runoff into streams and lakes. Local administration is under the supervision of an elected committee of farmers.

Such a federal program, dealing directly with farmers, could become the great innovator in enticing farmers to use the new technology as it becomes available for reducing pollution and the rate of eutrophication. Its beginning is small, one county and a limit of $2,500 to be paid to any one farm owner in a year.[10] Still, it is a promise and $2,500 is two-and-one-half percent per year of each $100,000 of capital investment in a farm, and it could make a bright difference.

The effort is significant in another sense. In the perpetual motion of farm politics and government public relations, a federal agency protective of farmers and a local committee of prominent farmers had admitted that farms did indeed contribute to the eutrophication of Lake Mendota and should do something about it.

ESTUARIES

While the oceans, even the continental shelves, are too vast a subject to be included in this discussion of water in the United States, they are certainly a part of the nation's environment. The shelves provide most of the fish caught and practically all the oil and minerals, including sand and gravel, now mined from the sea. The proper management of the use of continental shelves with a concern for environment is just as important as management on the land.[11]

The most urgent ocean water problem for the nation, however, is the estuary. Major examples of ocean estuaries are the lower Hudson River and the bays that make up the New York–New Jersey urban waterfronts, Chesapeake Bay, the Mississippi Delta, Galveston Bay, San Francisco Bay, and the mouth of the Columbia River. Until they become over-polluted, the estuaries provide recreation and edible fish, shell fish, and crustaceans. Some of them, notably Chesapeake and Galveston bays, are noted for the high quality of their oysters, crab, shrimp, and fish. One of the pleasant experiences of eating is to order such foods at a good seafood restaurant in Baltimore, Washington D.C., or Galveston.

All the estuaries are being polluted at a fatal rate. Much of the filth and chemicals dumped inland into rivers is emptied into the estuaries. Most of the waste from near-port shipping, the garbage and leaked oil, is dumped into the estuaries. In short, much of the pollution of streams that occurs inland reaches the estuaries.[12]

INVENTORIES ARE NEEDED

We know many more techniques for changing the quality of water than we have the will to use. This is usually the case in any question of environment.

The techniques vary in their state of refinement from one source to another. One of the dangers is that a technique that is only partially proved will be adopted and the polluter will then claim that he has reformed and met his obligations. He has reformed only part way. So the first step in the use of techniques is to find out how much pollution there is, where it comes from, and how much of it is diluted or degraded in water. A committee of chemists makes the point:

> Rational planning for water pollution control . . . requires detailed inventories of the composition and volume of all pollutants from all significant sources. Yet for only a few water systems, such as the Ohio River, the Delaware River, and Lake Erie have such inventories been compiled even partially. For almost no water pollutants, not excepting nutrients such as nitrogen-and phosphorus-containing compounds, have adequate balances been worked out of the amounts that enter the system and the amounts that leave. Gross inventories are available on the sources and amounts of municipal wastes, but data on industrial wastes are less readily available.

> The main inorganic constituents of most wastes include ions such as sodium, potassium, ammonium, calcium, magnesium, chloride, nitrate, nitrite, bicarbonate, sulfate, and phosphate. The specific organic compounds in *waterborne* wastes are less well known. Exceptions include the extensive programs of analysis for pesticides in surface waters, analytical work on synthetic detergents, and a few studies on phenolic substances and carboxylic acids in streams.

> Even the major organic chemical groups in domestic wastes and treated domestic wastes are known only partially. One of the relatively few analyses that have been made of domestic sewage could account for only 75% of the total organic carbon. The classes of compounds that were detected included carbohydrates, amino acids, fatty acids, soluble acids, esters, anionic surfactants, amino

sugars, amides, and creatinine. In other work, more than 40 specific compounds were identified in domestic sewage.

An analysis of the effluent from secondary sewage treatment could account for only about 35% of the total chemically oxidizable organic materials.[13]

Pollutants, whether from industries, from farms, or from municipalities, are essentially chemicals and minerals, and chemical–biochemical analysis is necessary before the technique of removal is indicated in a particular portion of water.

The methods for measuring the amount of pollution in water are chemical, but to understand them one must know what microscopic processes degrade pollutants. Again, the chemists know most:

> The processes that degrade and convert substances in water to other chemical and physical forms are extremely complicated because of the effects of aquatic life, mainly microorganisms. Microorganisms may control the soluble concentration of an element in water, notable examples being carbon, nitrogen, and phosphorus, the major elements in cells. Microorganisms may convert organic compounds in the water partly into carbon dioxide. They may convert dissolved carbon dioxide into organic compounds. They may affect the concentrations of inorganic compounds of silicon, aluminum, and other elements. Such effects are multiplied by the very large numbers of different substances in water and the variations in natural populations of microorganisms. Because of these complexities, general descriptions of biological degradation in water systems, particularly mathematical descriptions, seem unlikely to emerge. Research on degradation and its products must thus depend to a large extent on the measurement of specific substances in water, and their degradation products.[14]

Most commonly used for the measurement of pollution is Biochemical Oxygen Demand, always referred to among specialists as B.O.D. and written as BOD. BOD is a measurement of the weight of the dissolved oxygen used by microorganisms as they transform the carbon and nitrogen compounds in organic matter. Because the reactions necessary in the transformation take time during the period of incubation, a BOD test must cover a sufficient period of time. The standard BOD test covers five days. For the members of the Wisconsin Seminar who were not physical scientists, Edward E. Miller simplified BOD. "If a sample of water containing plenty of dissolved oxygen (and, of course, a natural inoculation of aerobic bacteria) is closed with no exposure to air for five days, the decrease of the oxygen concentration during that period (caused by the oxygen-burning consumption of organic matter accompanying the multiplication of bacteria) can be measured directly. This *decrease* of oxygen, measured in milligrams per liter of water (mg/l)—or in parts per million (ppm), essentially the same unit with a different name—is called the BOD. . . .

"When organic matter flows into a natural stream, the oxygen concentration of the stream drops off downstream—typically for 1 to 3 miles—as the bacterial action proceeds toward completion. Natural aeration thereafter begins to restore the oxygen again, returning the stream to a normal condition in perhaps a total of 2 to 10 miles—depending on a number of factors, of course. If the oxygen level approaches zero, the stream becomes anaerobic, and highly objectionable 'sewage odors' develop. For a given sewage concentration, the oxygen level can vary widely with stream flow, temperature, and other factors. Because of this variability, the dissolved oxygen present is a less meaningful and stable measure of the potential for sewage damage to aquatic communities than is the BOD.

"In the California publication 'Water Quality Criteria' it is noted that most states consider a BOD level of 2.5 or 3 mg/l (milligrams per liter of water) to be required of second class water—i.e. water suitable for swimming, fishing, and aesthetic enjoyment. I have therefore chosen 3 mg/l as a convenient standard or reference level for water quality. At less than this level there is no complaint; above this level we may begin to look for sources of trouble."

Professor Miller then proceeded to map the BOD measurements of streams in Dane and Iowa Counties and the sources of their pollution. He could obtain the measurements from studies already made by the State of Wisconsin Department of Natural Resources. Since the kinds of fish living in water differ according to the amount of oxygen available, Professor Miller also placed on the map the kinds of fish in the stream. He divided them by

Waste Sources and Quality of Fish, Dane County, Wisconsin

trout, which require the most oxygen; bass and/or pan fish, which rank next in the need of oxygen; and forage fish only, forage fish including minnows, buffalo, carp, and others which require the least oxygen. The data available were for 1963 and changes have been made since then in the sources of pollution. Nevertheless, the map for our two counties is a beautifully graphic presentation of the quality of water in the streams of these two counties.[15]

The same kind of inventory of quality can be mapped for any county or region in which the data have been collected for BOD and for the kinds of fish living in streams. In places where the data have not been collected, the mapping will take longer. The inventory of quality and sources of pollution must be available before the control of pollution can be thorough. Water pollution varies by many factors and by many pollutants, and sometimes the most innocent appearance will cover a serious threat. Handsome new rural homes have been found to have poor sewage disposal. And ugly debris may be relatively harmless in chemicals while it is a serious aesthetic pollutant.

Chemists accept BOD as only a partial measurement, although for most purposes acceptable. It fails to measure all the biochemical oxygen demand of the sample, missing some 20 percent, and it tells very little if anything about what happens to specific organic compounds that enter the water. Chemical Oxygen Demand, COD, "is based on the fact that most organic compounds can be oxidized to carbon dioxide and water by strong oxidizing agents. COD measures the equivalent oxygen demand of compounds that are biologically degradable and of many that are not. It thus gives higher values for oxygen demand than does the BOD test." A newer method of measurement is Total Organic Carbon, or TOC. "It is based on methods developed recently that involve rapid combustion of carbon and measurement of the resulting carbon dioxide by infrared spectroscopy." These two methods measure all the oxygen demand whenever such thoroughness is necessary. They still tell "little or nothing of the specific organic compounds in the sample of water that is analyzed." [16]

Whatever the method of measurement, it involves chemical

and biochemical analysis. Pollutants, again, are chemicals and minerals that get involved with living micro-organisms in the cleansing process of oxidation. If they reach a volume too great to allow oxygen to do its work in the time and distance available, the water becomes over-polluted.

There are, then, two kinds of inventory needed, one to show the amounts and sources of pollution, the other to show the chemical ingredients of the pollution.

One big exception to this chemical generality is heat as a pollutant. It comes into streams and lakes as warm or hot water discharged from factories and electric power plants, the major contributors, and it changes the ecology with results that are still largely unknown.

MORE INFORMATION IS NEEDED

It is impossible to generalize about the reduction of industrial waste. Some techniques will cost more than others. Some techniques will produce byproducts that will earn added money for the industry. Some chemicals can be allowed to enter the water in small amounts, and some will be harmless. Some will have to be barred altogether because they are deadly in a sufficient concentration and the only safe and sensible way to treat them is to say that none whatsoever may enter the water. Some of the "hard" pesticides—that is, those very slow to degrade—are in this category. And, to go back to the chemists' quotation, we need to know a lot more about the composition of both industrial and domestic waste before we can attain proper control.

"We need more research" is the haven for any committee, business, or government agency that wants to show concern but does not want to take action. It will be heard from many sides in the case of water pollution. Of course we need more research. We always need more research in any subject one can name. The recognition of a constant necessity for research has made basic and applied research one of our strongest, most thriving institutions.

MORE ACTION TOO

We need fast action too if American streams and lakes are to be saved for any use other than dumps and if the estuaries are to be given a chance to recover. We can start acting with the information we already have. As new information comes from research, we can change plans to include it or we can start whatever new programs may be indicated.

We already know that nitrates and phosphates speed up eutrophication. We already know that most streams are overloaded and never have a chance to cleanse their water. We know that estuaries are becoming unusable for fishing and recreation. We know that the sources of pollution are sewage plants, industrial plants, surface runoff, septic systems, municipalities that dump untreated sewage, and municipalities that treat their sewage inadequately. We know a lot more than we are putting into practice.

We can start by installing treatment plants in both industries and cities. Improvements can be made in methods of treatment. The plants will need to be changed as improved methods are discovered. But the plants can be installed now, at the present stage of technology. Most industries now have no facilities to treat waste, so they simply dump into city sewers instead, and industrial waste is more difficult to handle than ordinary sewage. Half the cities with more than 2,500 population do not have sanitary, nuisance-free disposal systems for sewage. The combining of sanitary sewers with storm sewers, which began a hundred years ago, meant in too many places that when a rain was heavy the treatment plant could not handle the overload. As a result, raw, untreated sewage was discharged and went into lakes and streams.[17] By now we have learned that storm sewers carry as serious pollutants as sanitary sewers, and the new answer seems to be that both kinds of sewage need treatment. Larger plants are indicated everywhere.

THE TREATMENT OF WASTE

The techniques of treatment are much the same for industry or for cities.

Generally (there are variations to any specific statement), sewage treatment involves two stages. Thus the terms "primary" and "secondary" plants have come into use to describe the extent to which cities or factories clean their wastes before releasing them into streams, lakes, and estuaries.

In the primary stage the sewage is screened to remove large solids, grit is removed, other solids may be ground up, and sludge is formed by allowing the sewage to clear while any solids that remain sink to the bottom of a tank to become sediment. If the primary stage is the only treatment, at this point the clear water above the sediment flows out and the sludge remains to be disposed of.

A secondary stage activates the sludge by aeration so that more oxygen can work on the materials in it. Sometimes bacteria are introduced into the aeration to transform organic matter into a harmless residue. After treatment, the activated sludge is allowed to settle so that heavy ingredients sink to the bottom, leaving clear water as the effluent, or the water that flows out of the treatment plant. (The word "effluent" is used almost solely to describe what comes out of a sewage treatment plant. It still means, however, that which flows out or flows forth.) In some cases the clear water, after the sediment has sunk to the bottom, is filtered through sand, and perhaps at the very end it is dosed with chlorine to kill any remaining bacteria.

Sludge remains in the plant after either the first or second stage. Disposing of this sludge is the most difficult and the most expensive part of sewage treatment. Any harmful organisms in it must be killed or controlled. The volume, still containing all the liquid, has to be reduced by drying out the liquid. Any useful ingredients in it should be recovered to be sold, although any large attempt to convert sludge into fertilizer has been difficult to make profitable. Various methods for doing all these things are in use. Except for the few cities that try to make the sale of fertilizer pay, the dry sludge is used as landfill or disposed of in lagoons or the ocean. If a sewage plant does not go through the process of drying the sludge, and some small plants do not, the sludge will be spread out on land to dry and be absorbed the the earth.

After second-stage treatment, sewage effluent still contains dissolved solids of minerals and chemicals, including the phosphate and nitrate which contribute most to lake eutrophication. Various technical ways of reducing these are either under test or in laboratory study, in a new field of research and testing of advanced, or third-stage, treatment of sewage. The results already promise that most of the remaining pollutants can be removed from effluent, including the removal of phosphates and nitrates.

Advanced treatment is practiced in the water reclamation plant of the South Tahoe Public Utility District in California. Elsewhere only a few modifications of truly advanced treatment have been tried.[18]

Drinking water in American cities would be periled if it were not purified before use. Streams are too often unpleasantly if not dangerously dirty. American lakes become more eutrophic by the day, all because sewage treatment largely falls short of what science and technology have already discovered to be possible.

FARMS AND MINES

We said in Chapter 2 that little is known about an economical and effective way of handling animal wastes on farms. The result is that surface runoff carries much of this sewage directly and untreated into streams. If industry and municipalities are mentioned as polluters of water and agriculture is not, it is only because so little is known about the extent of animal pollution. To sewage from animals and to human waste from faulty septic fields in rural areas should be added the fertilizers, herbicides, and pesticides which farmers spread on the soil.

Much of this chemical cover runs off with rain, snow, and melting ice into streams or lakes. Some of the same wastes will be found in surface runoff from a city, of course. The difference is that we know what to do about it in a city. We can treat storm sewer output just as we treat sanitary output. If storms overload a present combined treatment plant, we can build a larger plant. For new parts of a city we can build separate sewers and separate plants for storm and for sanitary sewage. On the farm we have not yet found a satisfactory way to handle all the millions of tons of animal excreta and surface runoff.

Plain soil is another primarily rural pollutant. Some of it is natural when it comes from terrain where man has made no changes. Most of it comes from the work of man and principally from the work of man as farmer. There is little excuse for any agricultural soil erosion in the United States. The techniques for controlling erosion, and for keeping valuable soil on the land, have been known and disseminated for years. They include small dams, contour plowing, cover crops, holding tanks, and pipe conduits. If a farmer cannot afford to invest the capital for improvement, the public should do it for him. A chunk of soil that breaks and falls into a stream in Minnesota is joined by others from elsewhere until no river is clear below Dubuque, Iowa. Many Americans are born, mature, and die without ever having seen a river whose water is clear of mud.

Finally, the drainage of mines, most of them abandoned and practically all of them coal mines, adds to water pollution. Iron sulfide or pyrite in a coal mine oxidizes to form sulfates, sulfuric acid, iron oxides, and other compounds. Water enters the mine and dissolves these products. The water now contains acidic solutions and eventually seeps into surface waters.

The treatment for this kind of pollution is prevention at the source. A mine can be sealed or flooded when it is abandoned so that air is prevented from starting the first step of oxidizing the iron sulfide or pyrite. If the mine is sealed, water would also be kept out. For water that is already contaminated from mine drainage, the solution is treatment, at present with lime, to neutralize the acid.

CONCLUSION ON THE EXTENT OF POLLUTION

To approach a conclusion about water pollution in the United States, one has to decide his own philosophy about the uses of streams, lakes, and estuaries. According to one view, streams, lakes, and estuaries should be used primarily as carriers and cesspools for waste. Usually this view includes a strictly dollar analysis of values. The income from the petrochemical industry along the Houston Ship Channel thus is more valuable than the income from the fish, oysters, shrimp, and crab still taken from Galveston Bay. In practice this attitude has produced in the Bay the equivalent daily contamination of raw sewage that would come from a city of 2 to 3 million people.[19] Someone who disposes of his waste in the Cuyahoga River in Ohio alike can argue

that his contribution to the public in taxes is much greater than
that of a commercial fishing firm before commercial fishing disappeared on Lake Erie. The price the public has to pay for his
product is less, he will say truly, if he does not have to clean the
waste before he dumps it into the river. Both the Houston Channel and the Cuyahoga River are far advanced examples of
streams used as sewers. One of the oddities that future historians
will find are news photographs of the Cuyahoga on fire because
the scum and debris on its surface were so thick it could burn.

- An argument based on immediate dollar costs will nearly always win. It has been winning since the settlement of Santa Fe,
Jamestown, and Plymouth.

The argument loses some weight if it is analyzed for long-range
dollar costs. Some of the pollutants injected into water must be
removed from water for use by people and by industry, and the
cost of purification is raised. Still the strictly dollar argument
will win. An increase in the cost of making water usable after
contamination will hardly offset the larger tax revenue and the
lower prices from business firms that pollute. Nor will the higher
cost of three-stage treatment of municipal sewage offset the saving in fees charged to users if the advanced treatment is not introduced.

All who are concerned for the total environment take the
other approach to pollution. They say that other costs than dollars are also important. The quality of life is involved. That neb-
ulous column called "social costs" should be added to the account. So far, attempts to put dollar figures on social costs have
been mainly incantations, but it does not matter. How does one
put a dollar cost on the life of a child? And why do it? What is
the dollar value of the pleasure derived from casting a fly in the
solitude of a clean stream, or in lying quietly beside still water?
What is the dollar value of the diversity so necessary to happiness and peace of mind?

We think that the preferences of men should be honored. For
lakes and estuaries these include boating, skiing, fishing, and
swimming; for streams, walking, fishing, sitting, and dreaming
dollarless dreams. They include the knowledge that wild creatures are also using the water and that we are not killing by pol-

lution the natural neighbors that we take pleasure in watching. We think that people have a right to clean water just as they have a right to make profits from a proper use of water as a resource. And we think that people should pay the higher dollar cost for keeping water clean.

From this point of view, the pollution of water in the United States has advanced to a state of abomination. Too many waters stink. Too many are glutted with debris. Too many are infested with the wrong bacteria and over-supplied with the wrong chemicals for safe and pleasant use. If pollution continues at the present rate of increase, most water in the nation soon will be unfit for any use except sewage.

Beyond man's pleasure, there is another argument for clean water. It is the unknown risk of waking some day to find that we have made the water unusable. All water recycles to be re-used. If man, as the only dirty animal, puts some permanent poison into water, perhaps a poison with a half-life of 50,000 years, he will have poisoned all the creatures that drink water and perhaps all plants as well. He can die thinking of the dollars he saved by not respecting the quality of water. He might think in that last moment that he should have been more prudent back in the last decades of the twentieth century.

The technical remedies for all pollution can be found in reducing by good management the causes of pollution. It is that simple.

THE SUPPLY OF WATER

We said earlier that there is plenty of water in the United States for the present population and technology, and probably for future growth to some uncertain extent. This is true only if the population, agriculture, and industry are distributed according to the supply of water. They are not, and so we hear frequent references to a shortage and we see in print with regularity the term "water crisis."

Where the shortage exists, it is due to two causes, usually both present at the same time. One is over-population in relation to supply. Over-population means too many people and too many industries in one spot, for in an urban-industrial society people

live by working in industry much more than by working on
farms. The other is the misuse of water so that it becomes difficult to use if not unusable.

The demand for water is impressive, but it varies greatly from
one class of user to another and from one kind of use to another
within a class. Roger Revelle, a geophysicist-oceanographer, describes this in tons:

> The amount of drinking water needed each year by human beings
> and domestic animals is of the order of 10 tons per ton of living
> tissue. Industrial requirements for washing, cooling and the circu-
> lation of materials range from one to two tons per ton of product
> in the manufacture of brick to 250 tons per ton of paper and 600
> tons per ton of nitrate fertilizer. Even the largest of these quanti-
> ties is small compared with the amounts of water needed in agri-
> culture. To grow a ton of sugar or corn under irrigation about
> 1,000 tons of water must be "consumed," that is, changed by soil
> evaporation and plant transpiration from liquid to vapor. Wheat,
> rice and cotton fiber respectively require about 1,500, 4,000 and
> 10,000 tons of water per ton of crop.[20]

It is undoubtedly too late to do anything about the existence
of Los Angeles and Phoenix, Southern California, and the arid
Southwest. They should never have been settled in the first place
by the number of people and industries that they now support
and add each year. In addition to people and industries has
come agriculture, including, in Arizona, cotton, which requires
10,000 tons of water for every ton of cotton produced. More than
one observer has noted wryly that Phoenix is located near the
site of a Hohokam Indian village that vanished sometime in the
fifteenth century after water in its irrigation system became so
salty it was unusable.

Modern Phoenix began to grow after adoption of the Reclamation Act in 1902. President Theodore Roosevelt, an expansive
man as well as a conservationist, thought it was a good idea to
reclaim arid land for agriculture by taking water from streams.
He was shortsighted, as so many advocates of irrigation have
been. By 1955 irrigation consumed about half of all the fresh
water consumed in the United States.[21]

Irrigation is a dangerous and deceptive practice. It offers

something for everybody. A big dam means big money for those
who build it. Water in a dry land is precious stuff; irrigation will
fill canals at low cost to the user. A lake above the dam brings a
welcome relief from aridity to cottage owners (and realtors), fishermen, boaters, and swimmers. Lake Mead in Nevada, behind
Hoover Dam, is a great attraction for visitors. But behind the
soothing promises lies the danger that urban populations and industry may grow beyond the limits of the available water and
new supplies will have to be found to support people and industries. And new acreage always can be opened to agriculture with
additional water.

In the end, solutions to the problem of over-population and
agriculture in deserts become very expensive, more costly than
sound economy should allow.

In some parts of the Southwest man is taking out ground
water from the earth faster than it is being replaced. Even the
present population will soon require new sources of supply. Hy-
drologists measure supply and demand for water in units related
to units of time: that is, by cubic feet per second, gallons per day,
and acre-feet per year, one acre-foot being the amount of water
required to cover one acre of land to a depth of one foot.

Mr. Revelle reports that the average supply of water in the
Southwest is 76 million acre-feet per year. If agriculture continued to grow at the 1963 rate, 98 to 131 million acre-feet would be
required by the year 2000 A.D. The deficit would have to be supplied by moving water over long distances. The capital cost
would be $30 to $70 billion; the annual cost, including payments
on the loan, would be $2 to $4 billion; the cost to users would be
$60 to $100 per acre-foot.

Such an end cost to users would be too high for most types of
agriculture. Only municipal, industrial, and recreational users
could afford it.

Southwestern agriculture developed contrary to the most economic use of water in the first place, says Nathaniel Wollman, in
a much quoted report prepared at the University of New Mexico. Water used in agriculture, he found, added to the economy
of the Southwest an average value of $44 to $51 per acre-foot.
The same water used for recreation would have added about

$250 per acre-foot. Used for industry the same water would have added from $3,000 to $4,000 per acre-foot.[22]

The urban complex of Southern California brings in water in huge pipes, as every school child knows, from lakes 240 miles away in the Sierra Nevada and from the Colorado River, 242 miles away. The area long ago grew beyond the limits of its ground water supply; it had, in fact, overdrawn them by pumping for irrigation, and the water level was lowered. The next big scheme proposed is to bring water south from the Feather River north of San Francisco.

If the Southwest and Los Angeles are examples of poor planning and misuse in arid settings, New York City is a good example of poor planning and misuse in a region of plentiful water. The State and the City allowed the Hudson River to become so polluted that its water now can be used only after such drastic treatment that it becomes distasteful. New York imports its water from reservoirs in hills far outside the city, and a long dry spell means a dangerous shortage for the City.

The misuse of water has already been illustrated by pollution. Other examples, discussed always with reference to particular places and practices, are waste of water wherever it is found, failure to reduce evaporation, practices that cause salt to develop in the ground water supply, and failure to take steps that would reduce the amount of water that soaks into the soil too quickly or that runs off before it is fully used. Techniques to correct these kinds of misuse are in practice in many places and are always under study.

Until man learns to make it rain exactly where he wants it to and in exactly the desired amount, he will have to do with his present supply of water. He can practice more conservation, and he can better plan his future needs and sources of water. The only other hope for significantly changing the supply of water is to transport it in huge quantities from one place to another. The most extensive use of transportation is overland through pipes or ditches and by redirecting the flow of streams. Overland transportation is used by New York and Los Angeles and for all systems of irrigation.

The most colossal scheme of all fills alike the dreams of conser-

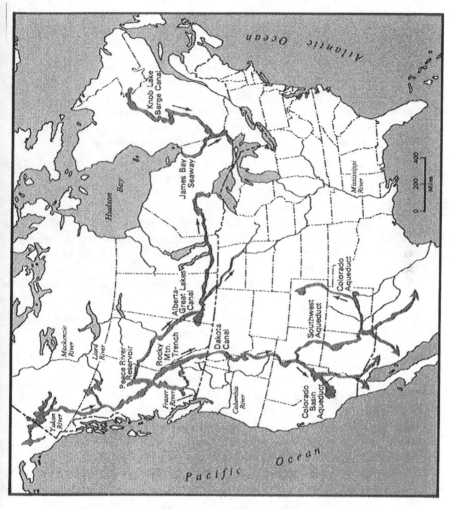

The biggest water diversion plan: Canada to Mexico.

95

vationists, users, and engineers. Their dreams are conditioned by what they forsee as the results. Those who advocate it will commit any offense against the land and nature in order to satisfy the increasing demand; they see in the dream a wonderful opportunity to really fix up nature on a truly grand scale. Those who know that nature will kick back in some unforeseen way, who believe in limiting demand by limiting population, who think that the nation should begin to plan where people concentrate by controlling water schemes and routes of transportation try their best to stop its development.

The scheme has more for everybody than just irrigation. It promises something to Canada, thirty-three American states, and Mexico. It offers fresh water from Alaska and the Canadian Northwest to the mountain and western states. It offers barge canals and inland seaways to Canada and all the Mississippi shippers. It cannot be grasped from words but only from a map, as presented here. Its advocates argue that it would mean as much to our future as the Louisiana Purchase. Its opponents ask what do we do for a second act? Where would we turn after the population doubles again? They advocate the control of consumers and consumption by limiting growth.

Plants that remove the salt from seawater and deliver freshwater already are used on ships and at stations on shore in several nations. In the United States the people of Freeport, Texas, use desalted water. A power company in Southern California has its own plant so that it will not draw from and pay for the local, imported water. The United States Naval Base at Guantanamo, Cuba, exists on salt-free seawater. Desalination was made necessary when Fidel Castro turned off Guantanamo's freshwater faucet in Cuba as part of his political quarrel with the United States.

The technical ability to desalt large amounts of seawater is no longer questioned. Salt can be removed from steam by a method known as flash distillation. It can be separated by electrolysis. It can be removed after separation from the water by freezing. Only the cost remains as a practical difficulty, and the cost promises to come down to a level that is economical.

Other possible consequences of desalination concern those who look ahead for the sake of environment. Luna B. Leopold, for one, points out that to desalt enough seawater to supply New York City for a year would leave a remaining pile of salt of 60 million tons, or more than is used in the United States in two years.[23] Of course, no advocates expect to pile such salt on land. When they put it back into the sea in such concentrated form, what will it do over a long range to the ecology of the sea plants and animals that are also useful? Then the usual concerns over thermal pollution, accidental radioactivity when the source of heat is nuclear, and unforeseen damage to the environment are raised. The argument settles again into a match between those who would satisfy demand no matter how it multiplies and those who favor controls to reduce demands, beginning with controls over the number of births.

NOTES

1. The most exhaustive source of information for laymen about water in general is United States Department of Agriculture, *The Yearbook of Agriculture 1955: Water* (Government Printing Office, Washington, D.C., 1955). Other useful, interesting, and readable books are Luna B. Leopold, Kenneth S. Davis, and the Editors of Life, *Water* (Time Inc., New York, 1966), Joseph A. Cocannouer, *Water and the Cycle of Life* (Devin-Adair, New York, 1958), Lorus and Margery Milne, *Water and Life* (Atheneum, New York, 1964), and Luna B. Leopold and Walter B. Langbein, *A Primer on Water* (Government Printing Office, Washington, D.C., 1960). The Wisconsin Seminar on Environment had available to it a paper on water by one of its members, Edward E. Miller, Professor of Physics and Soils.

2. Leopold *et al.*, *Water*, p. 38.

3. D.S. Halacy, Jr., *The Water Crisis* (E.P. Dutton, New York, 1966), p. 97.

4. Leopold *et al.*, *Water*, p. 177.

5. U.S. Department of Agriculture, *Yearbook*, *op. cit.*, article, "Underground Sources of Our Water" by Harold E. Thomas, pp. 62–77, especially the maps, pp. 66, 68, derived from maps in Harold E. Thomas, *The Conservation of Ground Water* (McGraw-Hill, New York, 1951).

6. Congressman Jim Wright, *The Coming Water Famine,* (Coward-McCann, New York, 1966), p. 140.

7. John Bardach, *Downstream: A Natural History of the River* (Harper & Row, New York, 1964), pp. 203–04.

8. The University of Wisconsin Water Resources Eutrophication Program, Madison, Wisconsin 53706, puts out a monthly series of abstracts of articles on eutrophication and adds a list of all documents collected by the Scientific Information Program in Eutrophication, at the same address. The most comprehensive discussion of eutrophication is *Eutrophication: Causes, Consequences, Correctives,* Proceedings of a Symposium held at the University of Wisconsin, June 11–15, 1967, under the sponsorship of the National Academy of Science–National Research Council, U.S. Atomic Energy Commission, U.S. Department of the Interior, National Science Foundation, and U.S. Office of Naval Research, (National Academy of Sciences, Washington, D.C., 1969).

9. Kathleen Sperry, "The Battle of Lake Erie: Eutrophication and Political Fragmentation," *Science,* Oct. 20, 1967; Sherwood Davidson Kohn, "Warning: The Green Slime is Here," *The New York Times Magazine,* March 22, 1970.

10. *Wisconsin State Journal,* Madison, Wisconsin, March 24, 1970.

11. K.O. Emery, "The Continental Shelves," *Scientific American,* Sept. 1969, p. 107.

12. For a good description of the near destruction of Galveston Bay, see "Galveston Bay: Test Case of an Estuary in Crisis," *Science,* Vol. 167, p. 1102, Feb. 20, 1970, a "News and Comment" article by Luther J. Carter.

13. *Cleaning Our Environment, The Chemical Basis for Action,* a report by the Subcommittee on Environmental Improvement, Committee on Chemistry and Public Affairs, American Chemical Society, Washington, D.C., 1969), p. 99.

14. *Ibid.,* p. 100.

15. Edward E. Miller, "Overview: Water in the Environment of Dane and Iowa Counties," mimeographed paper in the files of the Institute for Environmental Studies, University of Wisconsin, Madison, Wisconsin 53706. Only the map for Dane County is shown in the text.

16. *Cleaning Our Environment, op. cit.,* p. 101.

17. U.S. House of Representatives, *Environmental Pollution, A Challenge to Science and Technology,* Report of the Subcommittee on Science, Research, and Development of the Committee on Science and Astronautics (Government Printing Office, Washington, D.C., 1966), p. 38; *Cleaning Our Environment, op. cit.,* p. 139; Abel Wol-

man, "The Metabolism of Cities," *Scientific American*, Sept. 1965, pp. 184–85.

18. *Cleaning Our Environment, op. cit.*, pp. 106–38. This book is a thorough but concise discussion of the whole subject of the chemistry of air, water, solid wastes, and pesticides in the environment. The sources for all the statements are cited in the book.

19. "Galveston Bay," *Science, op. cit.*

20. Roger Revelle, "Water," *Scientific American*, Sept. 1963, p. 93.

21. *The Yearbook of Agriculture, 1955; Water, op. cit.*, p. 37.

22. *Ibid.*, pp. 95–96.

23. Leopold *et al.*, *Water*, p. 180.

chapter

—— 5 ——

THE ENVIRONMENT
OF THE SENSES

» » » » » » « « « « « «

MAN LIVES with more than land, air, and water as his environment. He also lives with sounds, odors, and lights. He lives with crowds and tempos.

The effects of these can be physiological or psychological, if these two can ever be separated, which is doubtful, but knowledge of what these effects are is sparse. Only a little is known about the physiological effects of noise, and less about odors and lights, while almost nothing is known about the effects in humans of crowds and tempos.

If it seems odd that so little is known about the consequences of the environment of the senses in a society where man is surrounded for hours each day with the sounds of motors, radios, records, and television, of whining jet planes and pounding machinery, of odors unlimited, of a confusion of pulsating lights and days of rush or monotony, it can be understood.

For one thing, facts about the most obvious parts of the environment were discovered to be important only recently. Until the sixth decade of the twentieth century man everywhere took the environment for granted. It was around him to be used. When it was discovered to be a threat, the discovery arose from the annoyance of smog, and only later was the effect of air pollution on health added to the account. Gradually, a few men began to point out that land and water were also significant and deteriorating in quality. We are just entering, in the seventh decade of the century, the stage of discovery when sounds, odors, lights, crowds, and tempos are considered to be part of the environment.

Another reason for the relative neglect of the senses is that physicians and physiologists, once they became interested in the senses, began with a preoccupation with the physical results of stimuli. They could seldom experiment with humans and so were restricted to animals, either small animals in laboratories or large animals in livestock husbandry. The medical definitions of senses were stressed. Studies were confined to how glands or blood pressure responded to stimuli and to how chemical reactions in the nerves were affected.

A good example of the interest in this kind of research was Walter Bradford Cannon. In 1906 he succeeded Henry P. Bowditch as George Higginson Professor of Physiology in the Harvard Medical School. The two of them were present at a beginning. Cannon wrote:

> The thirty-five years during which Dr. Bowditch held the chair . . . and the thirty-six years during which I held it covered the entire period of the development of physiology as an actively pursued medical science in the United States. Before 1871 it was a subject presented to the students in textbooks and in lectures commonly given by professors of medicine under the title "institutes of medicine." [1]

Professor Cannon published in 1915 what was to become a classic of medical research, his *Bodily Changes in Pain, Hunger, Fear, and Rage*. It was reissued in 1963.[2] He reported the results of experiments strictly in medical terms, without a hint that he was dealing with human emotions that would later be recognized as also a subject for medicine. But not easily. Psychosomatic medicine was still ignored by many physicians through the decades until 1960. Many medical schools dealt with emotions as aspects of neurology; physicians tended to assume that if they could not measure a change chemically or structurally, there was no explicable change. Until medicine and physiology began to recognize and deal with emotional states, there would be small interest in the human environment of the senses.

A third reason for the slowness to recognize the effects of the senses is the fact that such matters are hard to quantify. Scientists and engineers alike have long taken pride in dealing with *facts*, which to them meant countable facts; and if a thing could not be counted, measured, divided, and multiplied, they shunned it. Such things were known as unverifiable, the concern of philosophers and theologians.

This myth prevailed far longer than it should have; indeed it still prevails among the less educated, less thoughtful scientists and engineers. It was difficult to dispel because men did not recognize what was under their noses. They had inherited a system of measurement itself which was based on consensus, and there was nothing sacredly immutable about quantification. A foot was a foot, a meter a meter, because the people who measured accepted them as such. Yardsticks came from the minds and practices of men.

Today when panels of reasonable men reach an average conclusion about a standard of measurement, scientists and engineers are willing to accept this measurement, just as they accept feet, meters, pounds, or kilograms, which were born the same way.

An example is the decibel, when this unit to measure power is used to measure the level of sound. A decibel is partly "real" and partly "accepted by agreement" in the measurement of sound. It is real in that it measures the ratio between two pressures, one the pressure of sound. It is accepted in that the other pressure is an *agreed reference pressure*. Furthermore, the point of zero decibels is set at sound pressure that is just below the threshold of human hearing. Of course the ear's sensitivity to loudness will vary from one person to another and at different frequencies of sound.

To establish values, a "normal person" has to be found. The usual method is to collect a sample of persons who are known to be free of hearing defects and then to average the results of all their hearing. "This comparison has been done in great detail by many laboratories at various times and although minor discrepancies, some of them unexplained, have appeared the general agreement between results is good." [3]

A final reason for the slowness to recognize the environment of the senses is the difficulty of definition. In no other part of the total environment is there so much temptation to confuse personal tastes with facts. What constitutes visual noise? To one person the glare of lights that besets a driver is a danger; to another, it is an accepted condition that he does not think about. What is audible noise? Brahms' First Symphony, rather quiet music on the whole, can be played at high volume on a good high-fidelity system in a small room and sometimes register as high in decibels as a motorcycle five feet from the sidewalk. Is one noise and the other not? To the impervious young the overriding sound of rock music in a school common is not irritating; to their grandparents it can be maddening. Who is to define what odors are bad? One man's experience in a pulp-mill town has made him acquiescent to the smell; in the same town the next man's idealism makes him resentful.

When matters of the senses are so mixed with human values, scientists and engineers have been reluctant to assume that they are competent to define the good and the bad aspects of this part of the environment whenever the definition goes beyond the measurable. The tendency is, then, to define only those aspects of the environment of the senses from which damage to health is clearly tied to exposure to environment.

Thus, when deafness is clearly the result of industrial noise, sound above a certain level is said to be bad. Or when military sounds, especially in the Artillery and the Air Force, are found to be a threat to either hearing or efficiency, requirements will be set for wearing ear plugs or ear muffs. The noise of military and civilian jet airplanes after the measurement of levels of sound had begun was a clear threat to health and efficiency. Ground crews were given ear muffs. The same unarguable ties between environment and health have not been established for other parts of the environment of the senses.

"The external environment," wrote William Stumpf in a paper for the Wisconsin Seminar on Quality of Environment," can be described as an *energy surround* in which man is submerged, much as a fish in a pond. Man can be conceptualized as an *energy transducer* [something that converts input energy of one form into output energy of another form] in this surround, with limited sensing abilities of the energy forms such as temperature, pressure, electromagnetic radiation, and gravity. The human ear cannot perceive sounds above about 18,000 cycles per second; the eye is capable of transducing only very limited wavelengths of the electro magnetic spectrum. However, man's position on the phylogenetic scale suggests that he is unique from other animals, some of which possess superior sensing apparatus, because of his faculty to extract information from his energy surround, perform mediational functions, and respond in a wide range of adaptive behavior." [4]   Mr. Stumpf then asked, "What is the relationship between man and the environment?" and drew upon J. M. Fitch for a chart:

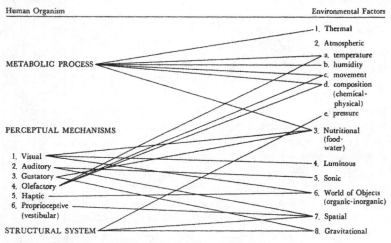

Relationship between Man and Environment

This chart is intended to illustrate the relationship between man and the energy surround. As J. M. Fitch has stated,

> The relationship of man's metabolic process to its environmental support is literally *uterine,* and since this process is the substructure or precondition to consciousness, sensory perception of changes in the environment in which the body finds itself is totally dependent upon the satisfaction of the body's minimal metabolic requirements. [5]

This is the relationship in which the physiologists and medical doctors have done their work. To get at the more difficult undefined and unmeasured question of the *quality* of environment for emotional satisfaction, for happiness or unhappiness, measurement has to be established. To establish measurement, man must first establish categories to be measured.

"A framework for measurement," continues Mr. Stumpf, "must be capable of relating controllable dimensions of the stimulus to behavior settings or functions. According to Berlyne these controllable variables include the following:

1. energy intensity (air, noise, light, etc.)
2. variation (temporal change)
3. complexity
4. novelty
5. surprisingness
6. incongruity

Ewald suggests others such as:

7. privacy
8. safety
9. density
10. number

and I might add one more, and doubtless there are many more,

11. escape

Ewald then proposes that these variables can be related to functions or behavior units listed below, using a matrix approach:

1. home
2. school
3. shopping
4. commuting
5. work
6. re-creation (spirit)

7. leisure and recreation
8. experience of nature
9. healing

"Assuming this type of oversimplified analysis is possible with the aid of computers, and that these subsystems of the larger urban and regional system can be optimized, it can be theorized that the summation of the subsystems would be one measure of the quality of the environment. . . .

"Until more comprehensive analysis and measurement are attainable, other ways must be investigated to create better guidelines for decisions (subjective though they may be) for use in the immediate future."

From this wonder of familiar complexity, we must choose those parts of the environment of the senses that can be discussed in this limited space. Some day, hopefully, much more will be known about the whole enormous and intimate system of man in his pond of energy and how much is pleasure and pain, his emotional states, his love and animosity are determined by his environment.

NOISE

Noise has many meanings. To an electronics engineer it is any disturbance that interferes with the reception of a signal. To a seismologist it is the steady slight agitation of the seismograph too small to indicate a significant disturbance of the earth's crust but too large to be accepted without allowance for it. To Mr. Stumpf: "Noise is defined . . . as any uncontrolled sensory stimuli in the environment that provides no useful information and that may in fact interfere with the reception of desired inputs." To an acoustics engineer noise is any sound that is not wanted by its recipient. To a physician noise is more likely to be any sound that is so loud it causes damage to physical or mental health. To one who is working on traffic safety noise is the distraction of drivers by flashing lights and too many signs. The common fact in all these definitions is that noise is disturbance and interference.

This kind of noise is undesirable. "It should be understood," adds Mr. Stumpf, "that some levels of noise . . . are desirable, that complete absence of background stimulation is undesirable."

We are concerned here with audible noise and, to the extent that any data can be found, with visual noise. Audible noise comes from traffic, airplanes, machinery, sirens, construction, amplified music, and speech. We will accept, and stick to, the definition of noise as sound that is unwanted by its recipient.

The difficulty of generalizing is apparent at once.

> A full-volume symphony may be music to an avowed musical enthusiast, but it will be noise to the neighbor trying to get her children to sleep. Similarly, the output of electric guitars may not be *physical* noise, but there are many parents of teenage children who might prefer to believe their psychological reaction than a spectrum analysis.[6]

The parents and their teenage children ought also to be thinking of physical noise and its physiological effects. The Committee on Environmental Quality of the Federal Council for Science and Technology reported in 1968:

> But annoyance is not the only piper we must pay. There is also a price in human health and efficiency. Prolonged exposure to intense noise produces permanent hearing loss. Increasing numbers of competent investigators believe that such exposure may adversely affect other organic, sensory and physiologic functions of the human body. . . . In short, growing numbers of researchers fear that the dangerous and hazardous effects of intense noise on human health are seriously underestimated.[7]

I am told by a man who has spent much time with hard rock bands that musicians know from experience that they can increase agitation in an audience by turning up the volume. Rock bands can buy special loudspeakers that will take high volume without "break-up." In the trade press these are advertised with brand names that suggest they are "fatal" to an audience.

Still, what is noise to some is entertainment to others, and any definition will be subject to question. Perhaps this is why most of the discussion of audible noise is confined to two kinds: (1) the noise that interferes with a designated function; for example, the noise of air conditioning in a concert hall; and (2) the noise that causes damage to health.

The definition of noise as sound that is unwanted by its recipient admits Brahms and rock, cheers and air hammers, or something as quiet as a dripping faucet in the night and something as loud as the boom of a nearby cannon.

Most of the concern with audible noise is, however, not with Brahms or faucets but with loud noises that either annoy people, or cause pain or deafness.

DEAFNESS

In the case of "directional sound," the level of sound varies with the distance of the source from the listener. And the quality of the sound, or whether it is an assault on the ear and the nervous system, varies with the pitch, or the cycles per second. A high-pitch sound is more "piercing" or "penetrating" than a low-pitch sound. The effect also varies with whether the sound is continuous or intermittent and whether the listener is exposed for a few minutes, a few hours, or days and years. The effect of noise can be temporary or permanent, or what the specialists call temporary or permanent "threshold shifts."

The human ear is, of course, a wonderful instrument, as are most human organs. It is a complex of fluids, bones, hair cells, membranes, and channels that can transmit sounds in all their great variety to the brain. It is also most sturdy, as most human organs are.

The ear is not sturdy enough, however, to take some of the urban-industrial noise. "Any of the moving parts . . . can be damaged or even destroyed by intense sound," says William Burns.[8]

Beyond this statement, that any moving part of the ear can be damaged by intense sound, it is impossible to generalize. One man's ear can take more or less than another's. Perhaps some idea of audible noise can be gained from listing the decibel counts of some directional sounds, with their distances, and of ambient sound in various environments.

First, some familiar levels are needed to allow ready comparisons. Conversational speech at three feet is between 60 and 70 decibels. A 10 horsepower outboard engine at 50 feet is just below 90 decibels. An automobile horn at 3 feet is between 110 and 120 decibels. A large pneumatic riveter at 4 feet is just under

130 decibels. A 50 horsepower siren at 100 feet is just under 140 decibels.

Some examples of ambient (surrounding) sounds are average traffic at 100 feet, just below 70 decibels, or inside a motor bus, 90 decibels. The complete list most often used of sources and loudness follows.

At a Given Distance from
Noise Source   *Environmental*

Decibels
RE 0.0002 MICROBAR

Threshold of pain	140	
50-HP Siren (100′)		
F-84 at Take-Off (80′ from Tail)		
Hydraulic Press (3′)	130	
Large Pneumatic Riveter (4′)		
		Boiler Shop (max. level)
Pneumatic Chipper (5′)		
Overhead 4-engine jet (500′)	120	Average Discomfort
Multiple Sand-Blast Unit (4′)		Engine Room of Submarine (full speed)
Trumpet Auto Horn (3′)		Jet Engine Test Control Room
Automatic Punch Press (3′)	110	
Construction compressors and Hammers (10′)		
Chipping Hammer (3′)		Woodworking Shop
Cut-Off Saw (2′)		Inside DC-6 Airliner
Annealing Furnace (4′)		Weaving Room
Rock Band (15′)	100	
Automatic Lathe (3′)		
Subway Train (20′)		Can Manufacturing Plant
Heavy Trucks (20′)		Inside Chicago Subway Car
Train Whistles (500′)	90	Inside Motor Bus
10-HP Outboard (50′)		Inside Sedan in City Traffic

Small Trucks Accelerating (30')		
Light Trucks in City (20')	80	Office with Tabulating Machines
Autos (20')		Heavy Traffic (25–50')
	70	
Conversational Speech (3')		Average Traffic (100') Accounting Office Chicago Industrial Areas
	60	
15,000 KVA, 115-KV Transformer (200')	50	Private Business Office Light Traffic (100') Average Residence
	40	
		Minimum Levels for Residential Areas in Chicago at Night
	30	Broadcasting Studio (Speech)
		Broadcasting Studio (Music)
	20	Studio for Sound Pictures
	10	
Threshold of Hearing— Young Men 1000 to 4000 CPS	0	

A level of average discomfort has been established at 120 decibels from 200 cycles per second to 10,000 cycles per second. The average threshold of pain to the listener is at 140 decibels from 200 cycles per second to 2,500 cycles per second.[9]

Deafness is some degree of deterioration in the threshold of hearing from some previously measured level. It can be measured only in a particular person. It can be temporary or permanent. It can develop gradually from exposure to too much sound, or it can come in a trauma from a short exposure or from one single loud sound. The most common traumatic deafness comes from a ruptured ear drum or damaged ossicles, the small bones of the inner ear.

The nature of deafness, especially the fact that it can be measured only in a particular person, makes it difficult to predict, and no satisfactory method of prediction has been found. Deafness varies in degree. It varies by the duration of exposure to loud sounds. It varies by the type of sound—for example, whether "piercing" or "full," although these terms among specialists are defined in terms of decibels related to cycles per second. It varies, of course, with the loudness of the sound.

Industrial noise causes the most concern because in many states damage to hearing entails payment of workmen's compensation for injury incurred on the job. (Wisconsin added deafness to its Workmen's Compensation law in 1958, the first state to do so). Factories, machines, air hammers and riveters, nearly any of the tools and processes used by industry, were first developed with little or no attention given to the amount of sound they would give off. After 1958 industries began to pay attention to noise, but not much, since no one could say for certain what level would bring on a claim of workmen's compensation after what length of exposure.

If industrial noise can cause deafness in those who work in noisy places or with noisy tools, community noise affects all who are surrounded by it. The noisiest contributors to community noise are airplanes and street or highway traffic. There is no firm evidence that community noise has caused any more than annoyance, perhaps because the incentive of workmen's compensation has not produced the studies.

When deafness cannot be predicted, the safest policy for private or public agencies is to set the permissible limits low. The United States Air Force began the practice in 1956, following data gathered by the American Standards Association. Its regulation stated that in the lower cycles per second a level of 85 decibels was probably not injurious, but that efforts should be made nonetheless to conserve hearing. This is about the level of small trucks accelerating 30 feet away, a 10-horsepower outboard engine 50 feet away, or inside a sedan in city traffic. A year later the American Academy of Opthalmology and Otolaryngology recommended protective measures at 85 decibels in the lower range of the scale.

Some specialists in deafness think that 85 decibels is too high for safety. If occupational noise is to be held at the level that surrounds urban workers not engaged in noisy occupations, says William Burns, the level would be at 65 or 70 decibles.[10] This is at the level of conversational speech at 3 feet, the Chicago industrial areas (the neighborhood), an accounting office, or average traffic at 100 feet.

A good rule of thumb to follow, if we accept this more cautious advice, is that *whenever noise is loud enough to cause persons three feet apart to raise their voices to be heard in ordinary conversation,* that noise may be damaging the hearing of all persons exposed consistently for eight hours a day, five days a week, through a working lifetime or perhaps an even shorter time.

ANNOYANCE

So far we have dealt with the physiological effects of noise in terms of damage to hearing. Another considerable effect is annoyance, usually defined as interference with sleep, speech, listening to radio or television, or any other disturbance that upsets the ease and tranquility of people who are not in control of the source of the noise.

Reactions vary greatly among individuals when this consequence is studied. We land again in the statistical morass of variable tastes, and what annoys one does not annoy another. Efforts have been made to connect nervous ailments and other traits of mental illness to annoyance by noise, but they have produced no firm evidence. We should point out, however, that studies of the effects of sonic booms caused by large supersonic airplanes are either so new or nonexistent that no conclusion is possible about them.

A public policy at this time to control noise as a source of annoyance would be based on social choice and not on matters of public health. This kind of policy, of course, is permissible and practiced by all governments. One coincidence should be mentioned. If the environment of the United States were made quiet enough to prevent deafness, it would also be quiet enough for everyone to enjoy life, whether or not they were annoyed by today's noise levels.

OTHER THAN NOISE

When we turn to the rest of the environment of the senses, we encounter a vacuum. Only assumptions support the accepted truths about life in the urban-industrial world. Crowds make the individual unhappy. Do they? There is no proof. It could be assumed just as easily that crowds stimulate individuals. For years the more imaginative young have escaped the small intimacies of farm and small town to seek the crowds in cities. Flashing lights make people nervous and lower their level of caution so that accidents happen. Do they? The opposite assumption is that flashing lights add to the interest of life. They break the monotony of sameness.

The bad odors of the industrial world are depressing and harmful. Yes, they are harmful when the odor indicates the presence of airborne chemicals that are harmful. But depressing? Prove first that people do not get used to them and ignore them. A young man who works summers in a paper pulp mill, surely a contender for man's most stinking creation, for it can be smelled downwind for miles, reports that after the first day he is accustomed to the stink and can eat a lunch of greasy sausage sandwiches in the agreeable company of other workers who have done the same thing for years.

Life today is too fast for good health and tranquility. Life is too complex for man to adjust. Life is too anonymous. Life is too Godless. Life is too materialistic. Life is simply ugly in this most advanced land of urban-industrialism. Saul Pett summed up this view in an eloquent indictment:

> Quality of life? you could start anywhere.
> With a new car that won't start or an old war that won't end or a dollar that won't stretch or an optimism that won't revive. Or a lake too dirty to swim in or a plane that is late or a supermarket checkout counter that resembles an exercise line for the catatonic. . . .
>
> We walk safely among the craters of the moon but not in the parks of New York or Chicago or Los Angeles. Technology and change run berserk. . . . The unthinkable multiplies. . . .

We live in an expanding theater of the absurd and the unreal. Between beers, we watch real men dying on television and, same station, same network, we get a poetic message about the dangers of smoking and a poetic message about the joys of smoking.[11]

Many Americans—and more foreigners—will agree with this description of life today in this complicated cockeyed land. They will not ask for data. They will assume that their preferences, or tastes, will be sufficient reason for concluding whether the quality of life is low.

The facts will not support them. René Dubos has earned eminence as a biological scientist who is concerned with man's quality of life. In the distinguished Silliman Lectures at Yale he concluded:

> The dangers posed by the agitation and tensions of modern life constitute another topic for which public fears are not based on valid evidence. . . . Indeed, there is no proof whatever that mental diseases are more common or more serious among city dwellers now than they were in the past. . . .

> The experience of urbanized and industrialized societies bears witness therefore to the fact that everywhere man can make some sort of adjustment to crowding, to environmental pollution, to emotional tensions, and certainly to many other kinds of organic and emotional stresses. . . . People of all origins, races, and colors, have now become adjusted to the artificial world created by modern industrial and urban civilization.[12]

Mr. Dubos, like some other biologists, makes a distinction between "the natural state" and the "artificial" urban-industrial technological environment. Some others think that the technological environment may be a part of the "natural state" of man, although they are not sure, because evolution in man cannot be measured in the short time he has been leaving records. Thus John McHale, an artist, designer, sociologist, and student of man's resources, writes:

> We have characterized [man's] technological development as the overlay of another evolutionary form on the natural genetic process. As there is little other comparative evidence, we do not know whether this overlay pattern of human extensions evolving for man may or may not be equally natural. . . .

The earliest uses of symbolic communication and tools mark a turning point when man became an active agent in his own development, when his species survival was no longer dependent on natural evolutionary processes. . . .

The early simple tools . . . which amplified the hitting and leverage power of the arm and hand, have now become complex assemblies of tools that amplify many-fold the combined limbs and hands of many men. . . .

We can trace this extension in many ways. From the skin as protective enclosure, we progress to clothes, houses, cars, planes, space capsules, and submarines as mobile coverings that give increasingly greater protection against environmental extremes. From the eye, we extend vision, and therefore survival advantage, through the microscope and telescope, the photo and television cameras, and on to sophisticated systems that record, amplify, and relate complex visual and aural patterns of great magnitude.[13]

MR. DUBOS ALSO WARNS

The very fact that man adapts to change in his environment carries threats. For one, the weak among us will be preserved and genetic deterioration can result because survival of the fittest ceases to operate as the way of genetic selection. For another, the physical damage accumulates slowly but shows up eventually in a high incidence of diseases from exposure to a hostile environment; for example, chronic bronchitis. Or "the possibility exists that the apparently successful adjustments to the emotional stresses caused by competitive behavior and crowding can result in delayed organic and mental disease." Unlike animals in nature, man, to be civilized, has to conceal his animal reactions to competition.

But the "most disturbing" penalty for man being so adaptable is that:

human beings . . . become adjusted to conditions and habits which will eventually destroy the values most characteristic of human life . . . they no longer mind the stench of automobile exhausts, or the ugliness generated by the urban sprawl; they regard it as normal to be trapped in automobile traffic, to spend much of a sunny afternoon on concrete highways among the dreariness of anonymous and amorphous streams of motor cars. Life in the modern city has become a symbol of the fact that man can become adapted to starless skies, treeless avenues, shapeless buildings, tasteless bread, joyless celebrations, spiritless pleasures—to a life without reverence for the past, love for the present, or hope for the future.

Man is so adaptable that he could survive and multiply in underground shelters, even though his regimented subterranean existence left him unaware of the robin's song in the spring, the whirl of dead leaves in the fall, and the moods of the wind—even though indeed all his ethical and esthetic values should wither.[14]

Human life, in other words, is more than biology, and a distinguished biologist is justified in deserting the certainty of the factual experiment when he speculates about human values, even when his values reflect his own sensitive taste. So far as we know, there are no statistics that show how many humans care how much for the robin's song in spring as compared with a safe regimented life underground. This does not matter. Man is human.

It is the very humanness of man that makes data from experiments and observations of animals only suggestive but not translatable to man.[15]

One bio-generalization about man and environment is safe and permissible. From his study of stress and endocrinology, Hans Selye could say:

No one can live without experiencing some degree of stress all the time. You may think that only serious disease or intensive physical or mental injury can cause stress. This is false. Crossing a busy intersection, exposure to a draft, or even sheer joy are enough to activate the body's stress-mechanism to some extent. Stress is not even necessarily bad for you; it is also the spice of life, for any emotion, any activity causes stress. But, of course, your system must be prepared to take it. The same stress which makes one person sick can be an invigorating experience for another. . . .

The secret of health and happiness lies in successful adjustment to the ever-changing conditions on this globe; the penalties for failure . . . are disease and unhappiness.[16]

THE RISKS MAN TAKES

To say that no firm data proves that physical damage comes to man from the environment of the senses is not to say that risks are not involved. We have already mentioned the possible genetic deterioration cited by René Dubos. The other risks come mainly from ignorance and indifference.

Until damage to the spirit is openly pathological, its scars do not show. The result is ignorance. Chronic bronchitis produces a cough, emphysema a shortness of breath. The poisons in food or soil cause visible illness when the amount becomes unendurable. Pollution of air and water can cause physical disturbances that range from irritation of the eyes to epidemic typhoid. But the drain on a nervous system that may come from assaults upon the senses is quiet and slow. When it does show up as a "nervous breakdown" or something worse, there is no way to know how much the cause might have been exposure to jarring lights and noises, to drab monotonous surroundings, and how much to such other factors as childhood, parents, fears, and similar acts of God.

One common definition of survival and health among living creatures is the condition which exists when the internal environment continues to adjust to changes in the external environment. The internal environment is where the internal cells have their being. So long as it remains in approximately the same state, no matter what is happening outside, the animal survives in good health. Unfortunately, the emotions do not readily show when an internal upset has occurred.

The knowledge man has of chemical changes under stress is limited almost entirely to knowledge of small animals. Because of man's social traits very little physiological research can be done on man himself. Certainly when the techniques of research call for killing the subject to perform an autopsy, man is exempt. Questions always remain about whether the biochemistry of rats can be applicable to man.

A similar knowledge blank is true for genetic change. The generations of man are far apart. Geneticists and other biologists, being men, cannot live long enough to trace change through several generations. They, and all the rest of us, must assume that what happens through quick generations of fruit flies suggests what happens in the genes of man.

Definitions are as unsettled for the results of stress as for the other components of happiness—unhappiness, tranquility—disturbance, or whatever words we use for the nervous and emotional consequences of the environment of the senses. Except for those advanced stages of collapse, such as schizophrenia, there is no accepted definition, and schizophrenia covers many varieties of behavior.

When definitions are uncertain, ignorance is common. A good example was a large study of the mental health of a sample of dwellers in midtown Manhattan published in 1962.[17] The daily press seized the story and announced that the study found 80 percent of midtown New Yorkers to have neurotic traits, 23 percent of them handicapped in their daily living.[18] Actually the report also showed many interesting facts about experiences, attitudes, and demographic traits of New Yorkers; but a close reading of the book, including the appendices, shows that the study of symptoms of neuroses was based on the authors' own definition of mental health and that they could find no standard definition. They used a psychiatrist's definition.[19] Anyone who has psychiatrists as friends will testify that they can see neurotic traits in the person who considers himself normal and who does his work without noticeable handicaps. A nervous stomach, the unexplained headache, tenseness before making a speech can all be called symptoms, but they do not interfere with work and life.

Finally, a lot of ignorance is due to a peculiar kind of social self-delusion that disables the professional intellectuals in American society. They think they know the answers when they don't. They answer one another so repeatedly with statements that are really no more than theories that they all begin to assume that the answers are correct. This kind of delusion comes regularly in America. Theory, hypothesis, becomes assumption which becomes fact, although at no phase of the development was any proof found. Answers are given like patellar reflexes in response to a Taylor percussion hammer.

"Drug addiction was caused by conditions in the slums and the hopelessness of the disfavored." This theory was taken as fact until physicians began to find addiction among the favored upper-income kids of suburbia. "An affluent suburb of New York like Fairfield County in Connecticut, which has the largest number of drug abusers in that state, is said to have as many as 10,000 addicts out of a total of 768,000 people." [20]

"Large families were the product of poverty and ignorance. If poor people knew how to use contraception and could pay the bill, the birthrate would drop until the threat of overpopulation would disappear." A "fact" assumed until the Census Bureau finally was heard by some to say that most American babies are born to middle-class parents who can pay for contraception and who practice it when they choose.

A good social scientist (some social scientists also practice delusion) tries to fight the ignorance caused by mistaking theory for fact, but he seldom wins. An error once widely accepted among the professionals becomes a property held jointly. To question it is to attack the verity of the group and the "knowledge" passed on to the young.

The examples are endless. Rural people are happier than city people because they drink less. (Drinking alcohol is not the only measure of happiness, and how much alcohol anyway?) The suicide rate among young people in New York's prisons is higher than anywhere (no evidence). New York City is a bad place to live; its suicide rate is higher than other places. (It has a higher rate of identifying suicide.) [21] All cities are bad places to live (no evidence). Small colleges are better for students than large ones (not proved).

Indifference toward the environment of the senses is one of those things that just happened without plan. It was at first part of a larger indifference toward the circumstance of man in the larger environment.

The indifference showed in the allocation of federal money for research. Through the 1950's the big push was for research in the physical sciences and especially in high-energy nuclear physics, a most expensive field. Through the 1960's the emphasis was laid upon biological sciences, and toward the end of the decade some attention was given to research on the environment. Still there was little money available for the research that was needed to study the psychological effects on man of factors in his environment. When research began, the emphasis was still on the biological results.

Social scientists themselves had not been interested in research on the human consequences of environment, again a case of unintended indifference. When Berelson and Steiner published in 1964 an inventory of research in human behavior, they made no reference to environment, stress, emotions in relation to environment, urban-rural differences in relation to environment, or annoyance.[22]

The result of indifference is that we know much more about the social behavior of mice in various orders of population than we do about people, more about the psychology of monkeys than of people, more about the behavior of birds than of people. The social scientists had had neither the money nor the interest to prepare the nation for racial outbursts, slum and black psychology, violence on the campuses, or for the possible psychology of the environment of the senses.

THE AVOIDANCE OF TOTALITY

Ignorance and indifference are due to the fact that there has been no science of human environment. Scholars have worked at pieces of many environments. Ecologists are botanists if they study plant environment or zoologists if they study animal environment. They are physiologists or physicians if they study the chemical consequences of stimuli from the environment of the senses or the cause and effect of deafness.

Promotion for the young men, additional merit for those who have already arrived, are won through specialized research. And the specialists and the specialties become narrower, more concentrated. An old mother science like physics breaks down into atomic theory, nuclear energy with high-energy and low-energy specialists, plasma physics, astrophysics, biophysics, geophysics, and others, until the specialists find it difficult to talk across the boundaries.

Only the artists have been concerned with the environment of the senses. Their concern has been with the effects of ugliness and monotony. They have found no hard facts from the natural and social sciences to help them. Recently a new breed of specialists in environmental design has appeared in a few universities. Some of them are trying to learn something about the sensory results of environment.

The truth is that the people in society who do the research and who administer the agencies where the research is done have avoided the total view. It is easier certainly to work on small, confinable parts. Perhaps it is impossible for one man to learn much about totality. The resort to teams of specialists is a result of this difficulty. Until someone does see totality as important, and acts on it, the consequences to man of his immersion in an environment of the senses will not be studied as a whole.

SOME QUESTIONS TO BE ASKED

About all that we can do to conclude this subject is to suggest some examples of questions that need answers. Other such questions will occur to other people.

Does artificial light have any effect on the physiology or emotions of man? Experiments with other creatures consistently show that life is geared to day and night, sunrise and sunset, light and dark. When this circadian rhythm, as it is called, is upset, changes begin to happen. Certain enzymes, the biochemical catalysts of body functions, will be over-produced or under-produced. Other functions that are geared to sunrise and sunset will adjust but only after a delay.

The crews of international air flights that cross several time zones within a day or night have learned to eat and sleep as far as possible on what they call stomach time, or the time at home. Astronauts live on earth home time. Government and business travelers are told to adjust to the new time on an international trip before they make any decisions or negotiate any agreements. Only the President appears to fly from one country to another and never stops talking. Possibly there is an explanation. The President has a bed on his plane and may catch some sleep in flight. Those little airport speeches usually say very little of consequence. It is still possible that a President may make a blooming big mistake if he makes a serious misstatement while unadjusted to his circadian rhythm.

What does artificial light in the profusion of the modern city do to men? We need to know.

City and inter-city highways surround man not only with artificial light but with busy light and a great variety of sounds, including a great deal of unwanted sound. Does a welter of lights and sounds put a strain on the nervous system or affect the emotions? Is the effect adverse?

Is there any connection between the profusion of signs, lights, and sounds and automobile safety? Or pedestrian safety? We have referred in this chapter to visual noise. Architects and some traffic engineers call this visual pollution, but they do so from common sense and not from facts.

In Madison the traffic engineers by rule of thumb say that no streetside advertising sign shall be more than thirty feet high and thus take a driver's eye from street level. No advertising sign shall resemble a sign that is used for traffic direction and warning. We have driven in other cities where an advertising sign is allowed to flash like police, fire, and ambulance warnings. It seems reasonable to propose that the time has come to spend money on research on signs, lights, sounds, and the strains associated with driving.

If urban man is made nervous by the stress of light, pulsation, and sound, will this condition become an element in natural selection until the genes carry the condition to the next generations, until the human species is changed over time by the condition in which man lives unaware today?

Man until the last seventy years lived almost entirely in a horizontal plane. Now in all the big cities he shifts to a vertical plane for several hours a day and for the whole day and night if his home is as high in a building as his office. Added variety occurs when his office or apartment has outer walls of glass so that his eye, when it turns downward, sees nothing but a drop to death. In one new Chicago building the tenants sometimes cannot see the street because of low-lying clouds. Does this distortion of a way of life, perhaps of an attachment to the earth, do things to man's nerves and emotions?

Most of all we need some statistical magic to tell us what is happening to man. The curves we have are very recent, beginning only when reporting was taken seriously as a public job. (Only the Census is long established; it was provided in the Constitution, 1789.) If there were some way to project backward, to reconstruct the facts of earlier times, we could say whether the present environment of the senses is more or less harmful than the environment of the past. It means little for trends to cite the number of hospital beds filled now by mental patients. Are there more mental cases now as a proportion of the population than a hundred years ago? Or ulcers, neuroses, and indigestion?

It may turn out that the environment of the senses is good for us. Or it may not affect us much one way or another. Or it may be causing damage now and for future generations. In this seventh decade of the second century of the industrial revolution we simply do not know.

NOTES

1. Walter Bradford Cannon, *Way of an Investigator, A Scientist's Experience in Medical Research* (Norton, New York, 1945), p. 20. By coincidence, Cannon was born, October 19, 1871, in Prairie du Chien, Wisconsin, which was earlier the site of Fort Crawford where William Beaumont, an Army surgeon, had one of the rare opportunities to do research on a human and where he "made his classic observations on digestion." Beaumont saved the life of and for years nursed a Canadian hunter, Alexis St. Martin, whose stomach had been permanently opened by a gunshot wound. St. Martin became known in the Wisconsin frontier of the 1820's as the man with a window in his stomach. Beaumont made observations through the window and became famous for his original contributions to the knowledge of digestion. *Ibid.*, pp. 13–14, 28–29.

2. (Harper & Row, New York, *Torchbook Series,* 1963).

3. P. H. Parkin and H. R. Humphreys, *Acoustics, Noise, and Buildings* (Frederick A. Praeger, New York, 1958), p. 33. For other explanations of the decibel as used to measure sound, see Don Davis, *Acoustical Tests and Measurements* (Bobbs-Merrill, Indianapolis, 1965), pp. 19–23; William Burns, *Noise and Man* (J. B. Lippincott,

Philadelphia, 1969), pp. 40–43; Andrew D. Hosey and Charles H. Powell (eds.), *Industrial Noise: A Guide to Its Evaluation and Control* (U.S. Public Health Service Publication No. 1572, Government Printing Office, Washington, D.C., 1967), pp. N–2–4 and N–2–5. A more sophisticated method of discovering normality of the auditory threshold, a distribution curve, and standard deviations is explained in Burns, *op. cit.,* pp. 80–82. The "average," however, is close enough.

4. William Stumpf, "Impact of Environment on Man: An Overview," mimeographed, Institute for Environmental Studies, The University of Wisconsin, Madison, Wisconsin 53706. All quotations not attributed to others are from this paper. Mr. Stumpf's use of the term "external environment" is due to the recognition among specialists of an internal environment in which the internal cells of the body live. Albert Szent-Györgyi, director of research at the Institute of Muscle Research, Woods Hole, Massachusetts, and winner of the Nobel Prize in Medicine in 1937, has suggested that man has a third environment in his spiritual self. "It looks as though we may ultimately master the pollution of the external environment, and the wonderful progress of medicine may take care of the internal one. So, our existence has become dependent on the third environment, the spiritual one." He finds the spiritual environment more heavily polluted than the external one—"by hatred, fear, jealousy, brutality, and rapacity; with the craving for domination and the shortsighted narrow national egotism and expansionism that could make our globe uninhabitable." See his "The Third Environment," *Saturday Review*, May 2, 1970, p. 63.

5. J. M. Fitch, "The Aesthetics of Function," *Annals of the New York Academy of Science,* 1963.

6. Michael Rodda, *Noise and Society* (Oliver and Boyd, London 1967).

7. Federal Council for Science and Technology, *Noise: Sound Without Value,* (Sept. 1968, processed), p. 3.

8. The latest, most comprehensive, easily available source on the effects of noise is William Burns, *Noise and Man, op. cit.* What follows is derived from this source.

9. Davis, *op. cit.,* pp. 8–9. Davis gives credit to General Radio Company. Federal Council for Science and Technology, *op. cit.*

10. Burns, *op. cit.,* p. 185.

11. Saul Pett, Associated Press, headlined "The High Noon of Our Discontent," in the Boston *Sunday Herald Traveler,* Feb. 15, 1970.

12. René Dubos, *Man Adapting* (Yale University Press, New Haven, 1965), p. 274.

13. John McHale, *The Future of the Future* (George Braziller, New York, 1969), pp. 98–99.

14. Dubos, *op. cit.*, pp. 276–79.

15. Dubos, *ibid.*, gives a well selected summary of various data about the effects of environmental factors on animals as well as data from safe experiments on man. Hans Selye, *The Stress of Life* (McGraw-Hill, New York, 1956), describes what happens to rats under stress caused by the injection of hormones and generalizes, with additional data from his experience with autopsies and observations on humans. One observation by Selye is:

> What makes me so certain that the natural human life span is far in excess of the actual one is this:
> Among all my autopsies (and I have performed quite a few), I have never seen a man who died of old age. . . . We invariably die because one vital part has worn out too early in proportion to the rest of the body (p. 276).

16. Selye, *ibid.*, from the Preface.

17. Leo Srole, Thomas S. Langner, Stanley T. Michael, Marvin K. Opler, Thomas A.C. Rennie, *Mental Health in the Metropolis: The Midtown Manhattan Study* (The Blakiston Division of McGraw-Hill, New York, 1962).

18. The source of this seems to be the following summary of symptoms in Table 8-3, p. 138, *ibid.*

> Home Survey Sample (Age 20–59), Respondents' Distribution on Symptom-formation Classification of Mental Health

Well	18.5%	
Mild Symptom Formation	36.3	
Moderate Symptom Formation	21.8	
Marked Symptom Formation	13.2	
Severe Symptom Formation	7.5	
Incapacitated Impair	2.7	
Impaired *		23.4
N = 100%	(1,660)	

* Marked, Severe, and Incapacitated [the last three items] combined.

19. *Ibid.*, Appendix F, pp. 395–407.

20. Lee Edson, "$C_{21} H_{23} NO_5$—A Primer for Parents and Children," *The New York Times Magazine*, May 24, 1970.

21. Harvey Swados, "The City's Island of the Damned," *The New York Times Magazine*, April 26, 1970.

22. Bernard Berelson and Gary A. Steiner, *Human Behavior, An Inventory of Scientific Findings* (Harcourt, Brace & World, New York, 1964). The headings in this book are an interesting list of what so-

cial scientists had been concerned with. They are *Perceiving, Learning and Thinking, Motivation, The Family, Face-to-Face Relations in Small Groups, Organization, Institutions, Social Stratification, Ethnic Relations, Mass Communication, Opinions-Attitudes-Beliefs, The Society,* and *Culture.*

chapter
6

THE AESTHETIC ENVIRONMENT

»»»»»»««««««

AESTHETICS MOVES the environment of the senses beyond nerves and glands or hearing into the definition of the beautiful and the ugly, and standards become even more difficult to define. We have to deal with the condition of man's soul when no two men seem to agree on what gives them spiritual comfort.

One first truth to establish is that the aesthetic environment is more than automobile graveyards and billboards. Because the subject in the United States seems to begin and end with these deposits on the landscape, the point must be made. (For some, to illustrate the difficulty of definition, automobile graveyards, these rude, undisciplined castoffs of the age, can be considered art, and the Guggenheim Museum has in its collection a smashed auto body painted in hues as beautiful as Joseph's coat.)

The environment as an aesthetic experience offers most prominently landscapes and cityscapes. It offers structures as units and as collections of units. It includes the physical surroundings of people in cities and suburbs, in the country, inside structures and inside vehicles.

Fred Logan, a member of the Wisconsin Seminar who has spent thirty years studying the city as aesthetic experience, and brooding over it, sums it up, "Aesthetic quality refers to the physical aspects . . . to which man reacts with an emotional feeling of satisfaction or dissatisfaction, induced by one or more of the senses." [1]

Philip H. Lewis, Jr., another member of the Seminar, in a report for the National Park Service of the rewards possible from regional planning, introduced as relevant to aesthetic pleasure, the four premises from which we started in Chapter 1:

1. The environment must offer the maximum possible in options of choice at any point in time.

2. The environment must offer the minimum possible of irreversible traits, and each decision made about its use has to avoid as much as possible the consequences of irreversibility.

3. . . . the environment must offer the maximum possible diversity.

4. . . . the good environment will allow men to grow in mind and spirit as well as allow them to stay in good physical health.

To Mr. Lewis, without any doubt, "Man has always sought change and variety in his search for life." [2]

SOME ACCEPTED TRUTHS

An agreement that diversity is essential for pleasure suggests that there may be other traditions, assumptions, "truths," that all those who talk about the aesthetic environment share. But before listing the accepted truths, we mention that a few specialists speak for beauty. Not all people are paying attention. Even when their lives are turned sour by ugliness, most people will not know the reason for it nor pay much heed to its possible cure. The artists and art critics, some of the architects and social philosophers, some of the city and regional planners, and a few nonprofessionals who are concerned are the defenders of beauty. Just as physicians and physiologists are the students of the environment of the senses, so these few are the spokesmen for beauty.

Whenever they are unanimous in accepting some article of faith, in discussing a subject where facts are either not possible or irrelevant, their agreement should offer guidelines to the engineers, contractors, managers, and legislators who change the physical environment every hour of every day in the United States. They accept not just one but several articles of faith. These recur in the books, articles, and conversations of those who are concerned for the aesthetic environment. Together these "truths" form the basis for standards of judgment and criticism when the use and appearance of the physical environment are discussed. They are all we have as guides. If there were scientific, objective conclusions about what is beautiful or ugly, what is monotonous, what gives pleasure through the senses, there would be no room for criticism of art, architecture, landscape, and urban design; and the whole field of philosophy which deals with aesthetics would not exist.

1. NATURE IS BEAUTIFUL

Fundamental to all aesthetic faith is the creed that nature is in itself beautiful and that man's sense of beauty has its source in man's relation to nature.

Our colleague Philip Lewis' talk, writing, and design is filled with this assumption, and he quotes the late Joseph Wood Krutch, the essayist, to make the point: "On some . . . country weekend we realize . . . that we have not merely escaped *from* something but have also entered *into* something; that we have joined the greatest of all communities, which is not that of men alone but of everything which shares, with man, the great adventure of being alive." [3]

Our colleague Fred Logan in his proposals for aesthetic relief in Madison assumed, without argument, that the lakes were assets of beauty and the view of them should not be impaired. People like the view of water; it gives them pleasure. We start from nature as beautiful. The occasional person who thinks lakes should be paved for parking lots is not typical of mankind. He is aesthetically depraved. [4]

Usually the reference to nature means landscapes, grass, trees, water, alternating hills and level lands, mountains and plains. But the word also admits, as Mr. Krutch does, the happenings in nature; all the life of other creatures is considered part of the pleasure of life.

2. HOLD NATURE

When man uses nature he should disturb it as little as possible. When he builds roads and houses, parking lots and airports, he should retain as much of the *look* and *function* of the site as possible. He should conform to the nature of the site or he will be inviting trouble.

The two purposes, one aesthetic and the other functional, can never be separated. If design works with nature, the use of the site will be more beautiful and also more economical. An eloquent spokesman for this point of view, Ian McHarg, opens his book with a chapter that shows how to plan the use of dunes and coastal brush, then describes the devastation from storms that came to the New Jersey shore because the nature of the dunes was ignored.[5] His solution saved as much of nature as was possible—it worked *with* nature, as he would say—and the result, if his solution had been used a hundred years ago, would have made the Jersey shore a beautiful place. In the long run, true economy also would have been served.

3. ACCENTUATE NATURE

When he cannot avoid disturbing nature, man should accentuate the nature that remains. This statement can almost be called a "natural law," it so pervades the talk and writing about environmental aesthetics. Thus Fred Logan can say, "The quality of visual environment is predetermined by the nature of the land, water, vegetation, and weather of its place on earth." He can go on from there to advocate steps that would use land in ways that would provide residents of Madison with views of all the lakes in the area. More lake frontage should be bought by the city. Street endings on the lakes should be turned into small parks. Only the lowest buildings should be allowed on the lake fronts, with a middle level next inland, and high rise buildings only in the center of the city's isthmus in the third tier inland from the lakes. Occupants of all three levels of buildings thus could have unrestricted views of the lakes. Yet streets, dwellings, and office buildings would have been admitted as necessary uses of the environment by man.

Man intrudes on nature, and uses nature, but he can, along with such use, accentuate the part of nature that remains.

Philip Lewis, by the same "natural law," could envision a natural region in south Madison, most of it already owned by the public. He could propose that man, in using all these acres, should tie together the remaining natural areas and accentuate them by creating paths and roadways where they were still needed. A nature trip through the city would start with open space and vistas on the University of Wisconsin campus, then proceed to a city park, through the University Arboretum, and then into an area where two farms have already been donated to the University, but where two additional farms would have to be bought. From here the "corridor" would include public land until the end of the trip in the center of the city, where the stately architecture of the finest domed capitol in the nation would add a final note. Blended with nature all the way, the man-made additions would complement the interest and beauty of the site.

A twelve-mile strip of nature and man's use of it, in a half circle nowhere more than five miles from the State Capitol at the center of town, would offer residents and visitors pleasant diversity. Similar "corridors," as Lewis calls them, can be found or planned deliberately for any city in the nation. This is the new landscape architecture. It accentuates nature in the use of land.

Such concern for nature is the beginning point for any architect, any critic, any official, any city planner, any citizen who has not lost all sense of beauty. It should be an element in all plans if the human amenities are to be saved. Public and private use, say the watchers of the aesthetic environment, should start with a concern for the beauty of nature.

A City Plan Commission introducing its proposals for central Madison, where declining business and residential property call for change, properly begins its report:

> Madison has a beautiful setting. Its natural attractiveness is fabled, and cannot escape even the most transient visitor. [This is a nostalgic view that city officials usually express. The present residents are often told that their city's beauty is fabled to one and all. It may have been at one time (Longfellow wrote a poem about the place after seeing two paintings of it by Thomas Moran of the Hudson River School) but not lately. Except for the lakes and capitol it has all the horrors of any other American city, and the lakes are usually too green to swim in and often stink.] The heart of its beauty lies in its isthmus, where the imposing Capitol dominates the downtown skyline.

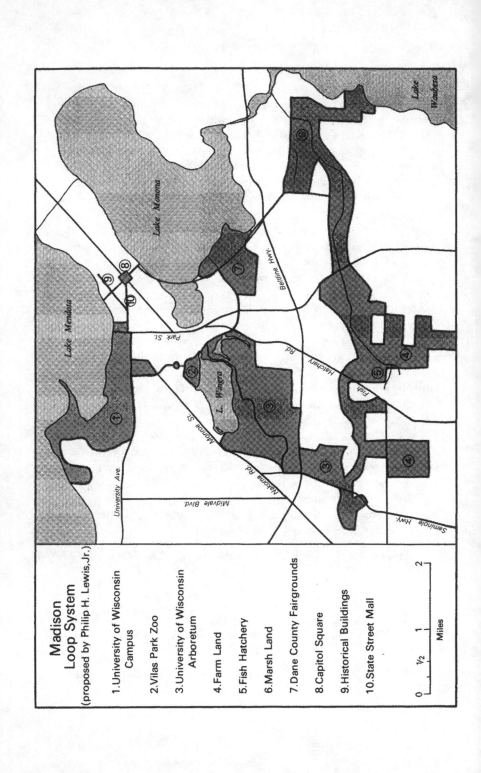

Madison
Loop System
(proposed by Philip H. Lewis, Jr.)

1. University of Wisconsin Campus
2. Vilas Park Zoo
3. University of Wisconsin Arboretum
4. Farm Land
5. Fish Hatchery
6. Marsh Land
7. Dane County Fairgrounds
8. Capitol Square
9. Historical Buildings
10. State Street Mall

Miles
0   ½   1   2

But Madison, like so many other cities, is concerned about the future of its central area:

Increasing numbers of cars . . . choke the narrow isthmus. This trend must be reversed. . . .

The central residential districts . . . must be redeveloped to provide an attractive environment not only to University students but also to those other Madisonians who would rather live downtown than in the suburbs, but who have no choice today.

Historic buildings are razed indiscriminately with other buildings, all in the name of progress. New buildings are often sterile and devoid of character. The best of the past must be saved, and the future instilled with vitality.[6]

The proposals that follow this introduction deal with streetscapes, meaning views of the Capitol, overhead wiring, trees, lights, historic preservation, design of buildings, and downtown functions and how they can be served with the most accentuation of natural beauty.

All the better city planners talk this way and wish they could make more of their insights come true. It is proper talk for planners with a fit sense of how nature should be accented in man's use of environment.

4. THE CITY CAN BE BEAUTIFUL

The city as a way of life is so old that it is part of man's natural history and therefore a part of beauty in the environment when it is allowed to be. Nothing inherent in the city forbids it's being a beautiful place to live.

This needs to be remembered in recent times. So many things have gone wrong and been left undone in urban development. Most observers' pleasure with occasional tastefully designed buildings is tempered by the thought of the ugliness of so many thousands of others.

When an attractive building appears, it may represent a retreat from the surrounding city. Thus the Ford Foundation building in New York, which won praise from most critics, is beautiful because it contains within a glass enclosed court the natural look of tree and bush and water. A person working in this building can look down from a balcony into the most expensive backyard in the city, a 10-story enclosed plot where a tree grows. To see other trees and bushes, the same person would have to walk to a park.

A step outside the Ford Foundation building puts one on East 43rd Street, to breathe air made dangerous by smoke and internal combustion. Sidewalks are littered. Breezes raise dust and dirty bits of paper. A few blocks away, to break the monotony of dirt, the United Nations buildings and grounds provide another point of visual interest against the background of the East River.

The city grew from man's social nature and the necessities of family and tribal living. Lewis Mumford, the philosopher who has written most about the city, speculates that the city grew from camps for hunting and collecting. Man lived in these village-like camps because his culture dictated that he be near his dead and he needed sanctuary. From the beginning there must have been a secure feeling about living near other men. Indeed the pleasure of living inside a warm and dry structure may have started very early. "Not least, perhaps, the cave gave early man his first conception of architectural space, his first glimpse of the power of a walled enclosure to intensify spiritual receptivity and emotional exaltation," says Mr. Mumford.[7]

The city can be beautiful, but few are; and those few are beautiful only in certain parts, from certain perspectives, or in certain aspects. Visually, the Borough of Manhattan is the most beautiful city in the world. From street level its buildings give the pedestrian every conceivable emotional experience: excitement, the quiet pleasure of brownstone houses, the elegance of magnificent store windows, the variety of districts where types of business are concentrated in the nearest thing we have to communities of trade and finance. From the bay or from the East and West Side highways, the skyline is breathtaking on a clear day. Our most eminent art critic, John Canaday of *The New York Times,* says, "Whether they are ugly or beautiful, the individual buildings of New York City are lost in the spectacle of its skyline, the crowning architectural achievement of the twentieth century."[8]

None of this was designed. Geography and the necessity to provide millions of people with places to live and work meant that land on this unique island would become more costly than perhaps any other land on earth. The telephone was invented, which meant that these millions of people could transact their affairs without ever meeting face to face. The electric elevator meant that they could rise to great heights inside buildings. So the high buildings were built to produce more revenue for each square foot of real estate.

It happened that the headquarters for banking, brokerage, and shipping remained in their old place at the foot of the island, and their profiles became the lower Manhattan skyline. Hotels, publishing, radio and television, and the newer industries such as business machines, automobiles, chemicals, and airlines added high buildings in midtown.

If the visual pleasure of Manhattan's streets and skyline could be enjoyed without dirt, noise, tension, traffic jams, and the leaden fear of breathing polluted air, Manhattan would be an aesthetic paradise. All other American cities share these disadvantages. The difference between them and Manhattan is that none of the others has as many splendid buildings as Manhattan, and none has the stimulation that comes from the sheer weight and height of the unique skyline.

Some other American cities retain beauty from their past. Washington still shows that it was a planned city. San Francisco in its heart displays a special charm, partly geographic, partly from the spirit of the people on the streets. Chicago's Michigan Avenue is one of the grand streets of the world, and its Lake Front with Grant Park and the Outer Drive make good entrances and exits to and from the city center. The central district of Minneapolis has the excitement and appearance of metropolis. Anyone who has ever visited Santa Fe and stayed to become acquainted with it will never forget it. Madison, from the air, is beautiful.

But in all these cities, a person looking for aesthetic satisfaction must search out the few vistas, blocks, the street-endings, or the heights from which the city looks and feels beautiful. And always, to enjoy the city, he must get used to stress and pollution, or overlook them.

All the cities, including the other boroughs of New York, become expanses of monotony in their newer sections. Miles of dwellings, shops, and factories have been built in the architectural fads of their times and add up to nothing of interest. Thoroughfares have been widened to carry ever more automobiles to and from the center of town. An arterial thoroughfare in any city in America has lost all aesthetic meaning. It is an open sewer of traffic polluted with noise, carbon monoxide, and lead. A child may not safely play beside this road nor a grown man walk.

The newest thoroughfare, the freeway, kills what might be saved of urban pleasures. It pours cars into the central city by day and drains them out at night. It kills neighborhoods when it cuts through them. It ruins forever any chance a person might once have had to walk along it or across it. The freeway is official statement that nearly all citizens have automobiles and the first purpose of the city is to move automobiles, urban amenities notwithstanding.

The better architects and designers dream a lot when they look at cities. They dream sensibly and humanely. They draw up plans to show how urban life can be improved—how automobile traffic can be separated from pedestrians; how shopping malls can provide quiet retreats from the rush; how mass transit trains can serve the metropolis, leaving private automobiles for pleasure driving in the country and for trips; how parked cars can be kept out of sight if they are allowed in the center of the city at all; how slums can be renovated; how landscape design can accentuate nature within the city, with grass and trees in areas surrounded by homes.

Only occasionally does a dream come true, however, and never on the full scale of a city redone. The *status quo* usually wins. The businessmen who operate walk-in stores are ignorant and fearful of change. They think that automobile traffic and streetside parking are necessary to get customers into their stores. The owners of real estate, and speculators, know the present and want none of the risk a different future might bring. Government itself may lose from change if it collects tolls from parking meters or bridges and tunnels. Any threat of higher taxes to pay for change turns taxpayers into protesters.

Designers know all this. When they try to regain or preserve some aesthetic values in their plans, they come face to face with reality and thus talk most about escape. All that remains possible for the old central city, given the robot brutality of the automobile and man's choice to depend on it, is to plan different ways to handle traffic.

The model cities since early in the century have really been model escape suburbs, although some of them were conceived with the idea that they would be self-sufficient and that their residents would find work in them. Actually suburbs of New York, Washington, Milwaukee, Cincinnati, and Baltimore, the first were Radburn, New Jersey, and Sunnyside Garden, Long Island, built in the 1920's; Greenbelt, Maryland, Greendale, Wisconsin, and Greenhills, Ohio, in the 1930's; and Reston, Virginia, and Columbia, Maryland, in the 1960's.

Now there is talk that future planning should be based strictly on science. Since the city can be considered an organism, with a metabolic process that takes in people and cars, water and goods, and excretes people, cars, and wastes, a city calls more for science and engineering than for architectural design. If the model cities of the past started from economic and aesthetic concerns, the model city of the future will be founded in science.

Escape to the suburbs, new and model though they might be in a very few cases, does little, however, to make the city itself a decent place in which to live. The really hard work of sprucing up the aesthetic aspect of old cities is scarcely considered.

5. THE PAST MAKES THE PRESENT

John Canaday finds the great themes of art to be the creation, nature, work and daily life, war and peace, diversions and games, still-life, "the spirit behind the mask" of a face, and dream and fantasy.[9] He is talking about art as defined by critics, dealers, and curators: painting and sculpture, usually found indoors, since only a few American cities have outdoor sculpture worth a second look. Some other observers would add interior and industrial design and crafts.

Some of these themes have been around since man began to leave records. They are revealed in the Lascaux cave paintings of

15,000 years ago and the Dordogne sculptures. They are reflected in Stonehenge and the megaliths of Brittany. Styles and techniques change but the themes do not. The shape and look of the present environment is determined by the past.

Paul D. Spreiregen of the American Institute of Architects suggests that the invention of the plow determined the rectilinear shape of later cities. The plow made straight lines. Some plan for dividing land and assigning it for cultivation was needed, and for reapportionment after the annual floods of the Tigris, Euphrates, and Nile rivers beside which, as far as we know now, the first major cities grew. The logic of lines was transferred to the cities for the same reason—to divide real property. Streets were straight, as straight as possible; gardens and ceremonial courts were square or rectangular; mud bricks and the houses made from them were square or rectangular.

But some early cities, and some sites within square cities, were circular. These, Mr. Spreiregen thinks, came first from the herdsman, who descended from the hunter and who preceded the warrior. The circle was best for fencing in a herd. It enclosed the most land with the least fencing. It also kept enemies out; the chief use of the circular town was to come later for defense.[10]

The most extensive deliberate use of straight lines to divide property was in the Northwest Territory of the United States, the vast area that included land which later became some of the States that touch the Great Lakes. This land came under the jurisdiction of the national government during the twelve years of the Articles of Confederation, before the Constitution was adopted. The territory was surveyed by townships of 36 square miles each, and these were divided into sections of 1 square mile each, or 640 acres. Townships were divided into farms and cities, as cities were added. The unbroken lineation that resulted can be seen from any airplane flying west of Pittsburgh on a clear day. Farm roads and city streets conform to the section lines. Only rivers, transcontinental highways, city freeways, and railroads make diagonal or curving lines.

The relentless grid in one city of the Northwest Territory, Madison. Streets in the beginning were section lines.

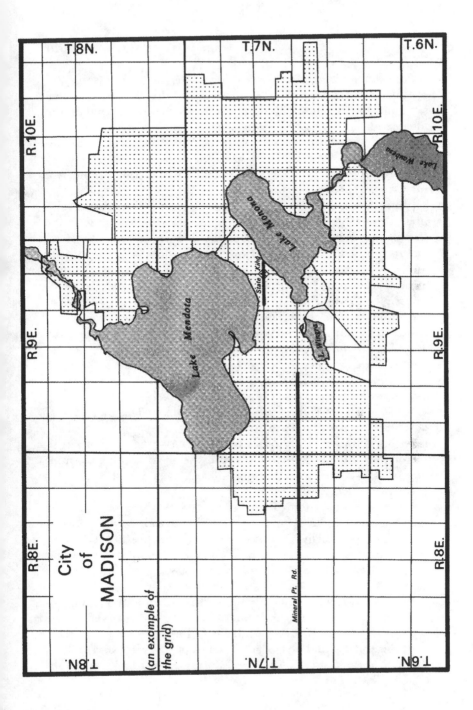

For one city in the former Northwest Territory, Madison, Wisconsin, Fred Logan finds, "The survey [of the city] was inevitably a product of the regulations of the Northwest Ordinance, the nineteenth century admiration of order and efficiency, and the universal acceptance of the plan of Washington D.C., as the correct setting for new state governments. The township and sectional lines were the basic surveys made by federally authorized surveyors.

"King Street-State Street is laid out on an east-west township line. Mineral Point Road is the next east-west section line to the south. The westernmost point of the first survey ended exactly one section width (one mile) from the Capitol at a point roughly on the southeastern corner of Bascom Hall [on the University of Wisconsin campus].

"Obviously the diagonal streets radiating from the corners of the Square and superimposed on the gridiron of the isthmus constitute the minimal effort to imitate the Washington plan by Major L'Enfant. Finally, the repetition of identical-sized lots in identical-sized blocks was deemed an orderly pattern for town movement and also permitted ease of speculative sale and resale of real estate. . . .

"One other decision . . . was the irregular extension of Langdon Street to parallel State Street. All other lots, regardless of the lake shores, ran northwest to southeast. Doty [the owner-speculator of the original city site] reserved the north-south lake front lots on Langdon Street for himself and men important to his plans in the legislature."

From a combination of respect for the Washington plan, the greed of a real estate speculator, and his need for favors to barter with venal politicians, Madison made a small break with linearity. A street was bent to provide some choicer lots. Most other cities of the Northwest Territory were not so fortunate. They are as rectilinear as checkerboards.

Later Doty won in his scheme to have the State Capitol located in Madison, to be built on a square he had reserved on the highest hill in the isthmus. Still later the University was given land on and beyond the edge of town, on the first high hill to the west of the Capitol. These were fortunate beginning possibilities for aesthetic development. Like most cities Madison wasted its heritage.

6. REMINDERS OF THE PAST ARE VALUABLE

To the thoughtful natural scientist some of the past is valued because it provides sites for research and it may be needed in some unforeseen development. The past is one of the options that should be available to choose in the next step of change.

The thoughtful observer of aesthetics values the past for other reasons. He feels just as concerned as the natural scientist. Many of the works from the past are beautiful in themselves. Taste changes from century to century and within a century. Still, there will be little argument about the great works of art, the sculpture, paintings, and buildings that have been preserved since the early great days of Egypt and Assyria. If classical art had been tossed out as trash, the earth would have been an uglier planet than it is today. There is about this art a continuity and agreement about what makes it great, from the prehistoric painters of Lascaux to the works on display in the Museum of Modern Art.

The instinct to preserve more recent and less authenticated art objects of the past is different. Sometimes the work reappears, highly esteemed, in a later generation. One trivial example is Tiffany glass, especially when used in lamps. This pleasant furnishing was used widely in the early 1900's then disappeared into attics or dump heaps. In the 1960's it became fashionable again.

More interesting are resurrected architectural fashions. Homes built in colonial, ranch, or Cape Cod fashion are copies of the past. Recently, a new appreciation has developed for the steamboat gothic house, also known as carpenter gothic and ginger bread. Russell Lynes, the historian of American taste, credits Andrew Jackson Downing, of Newburgh, New York, with exerting a strong influence on taste and having a "profound impact on the looks of the countryside." Downing, a landscape architect, promoted the rolling, carefully trimmed lawn. He tied the appearance of country homes to the state of civilization. "So long as men are forced to dwell in log huts and follow the hunter's life, we must not be surprised at lynch law and the use of the bowie knife," he wrote. Order and culture are reflected in "smiling lawns and tasteful cottages."

The kind of tasteful cottage that Downing built for himself and recommended for all his countrymen was a simple Gothic Revival. He thought the Greek Revival dwelling a "tasteless temple." [11] The kind of house that Downing favored was not as fancy as its descendant. It did, however, provide the base from which giddy carpenters took off. It had bay windows and turrets, and its roof was steep and gabled. The scroll saw was used after Downing to ornament the edges of roofs or the columns of porches with some of the fanciest wood decorations ever seen in any land.

After the fashion changed, these houses for years were nothing but sources of embarassment to their owners. Then, during the middle years of the twentieth century, they became charming, rather noble items for collectors.

Other reasons can be given for preserving reminders of the past. They allow comparisons with other cultures. African sculpture was around for centuries, but was never really appreciated until someone saw its kinship with modern European art. Someday the preColumbian art of Central and South America may give clues to other kinships. As more is learned about the Olmecs, who originated the preColumbian culture in southern Mexico, more will be known about the ties between the huge heads they sculptured and their social and religious system.

The past can always be a source of ideas for the present. Moreover, if the art was good in its day, it is good in our day. A small town in the Midwest that has a Louis Sullivan bank, designed in his latter years by our greatest architect, has a precious possession. The one I know, and go to see, is in Columbus, Wisconsin, 18 miles from Madison. The plentiful ornamentation speaks of a period when there was enough money and time to make a building beautiful simply for the sake of beauty. And that is a comfort to see today in a society that thinks it is affluent but buys less beauty than its predecessors bought.

The best argument for preserving reminders of the past is no argument at all. Old things have a beauty of their own. They are different. In our society they represent a more plentiful time, when good material was used and craftsmanship was leisurely but thorough. They are interesting. For those who pay attention to the aesthetic environment, reminders of the past are valuable simply because they exist.

7. POPULAR TASTE IS UNRELIABLE

The taste of mass consumers, the sponsors of the good aesthetic environment agree, cannot be trusted to ensure environmental beauty. This taste has been forced to a low-level standard of conformity by propaganda and mass entertainment. It endorses the ranch style home for all parts of the country, as if all places were as appropriate for it as the spacious rural Southwest, where the style was first adopted. In Los Angeles where air pollution for years has exceeded danger, it chooses to vote more money to build more freeways instead of improving mass transit. It is insensitive to streets made ugly by signs and overhead wires or to automobiles in the center of town that make life unpleasant and unhealthy.

To give the public what it wants in the aesthetic environment is to give the public what it has been told to want, and that is not determined by what makes life most pleasant.

Then who contributes what beauty we have among the works of man? The spokesmen for beauty do, the artists, architects, critics, designers, philosophers, and the other concerned people. Beauty is the work of a few individuals.

8. BLIGHTS

It follows that environmental blights are worst where the accepted truths about aesthetics are least honored. A city that allows a historic house to be destroyed will soon have the appearance everywhere of drive-in desolation. A state that has no respect for nature will end up as a huge washed-out gulley and with a social morale that is born in ugliness and reduced to shame.

9. PREVENTION IS BEST

The easiest way out is to prevent further damage to the aesthetic environment. Many journeyman planners, who also have souls, can instruct business executives, home builders, and public officials how to get the most beauty possible from nature and how to secure all other items in the creed just offered.

10. REDEMPTION IS UNLIKELY

Redemption is harder. It means undoing some of the property rights involved, perhaps, or persuading people to change their ways. Since most changes in the environment come from the actions of people, change usually reflects social traits. And social traits are the most inflexible, most durable aspects of society. To go back to Chapter 1, nature will renew itself if allowed to, but once a society has adopted a way of behavior, it may be impossible to change it.

Obviously prevention is recommended, since redemption is so unlikely.

MATTERS OF THE SOUL

Among all the creatures of the earth, only man has a soul. Beauty is a pleasure to the soul. Ugliness harms the soul. It is this simple. The statement has to be accepted as a universal truth and not one held only by the special spokesmen for beauty. Aesthetic pleasure raises the spirit of man and makes him better than he might be without it.

Andrew Jackson Downing was too narrow in his definition of aesthetic uplift—more than smiling lawns and tasteful cottages are needed for the good life—but he was right in his main contention: as long as men are forced to live in brutal conditions, we can expect from them brutal behavior.

This point comes up in every discussion of crime in the cities and drugs in the suburbs. Humans are presumed to turn to crime, dope, and drink because they cannot find fulfillment in the grind of their everyday lives. Monotony is a denial of beauty. So is boredom. Persistent frustration suggests that humans in their souls are defeated by their environment, whether it be rats and decay in the ghettos or the aesthetic emptiness of the architecture in urban fringe and suburbs. Implicit in all reform is the assumption that life can be made more pleasant by improving man's visual, auditory, and sensual surroundings, that men ought to be treated more like men and less like animals, since, after all, men have souls and can appreciate the nobler things.

Looking around America as the twentieth century ebbs, we see that man has defiled and ignored beauty for so long that we must now begin to ask why. Man is said to be the most intelligent of all creatures. He is said to be the only one with a soul and an appreciation for beauty. He was created by the very God Himself and placed beautifully naked and innocent on earth, which he proceeded to defile and has continued to defile through all his time here.

The scientific answers about man's insensitivity to beauty will come, if ever, only from profound research into man's psychology and social history. So far the social scientists have shown almost no interest in searching for the reasons for man's tolerance of ugliness, nor have the members of any other academic discipline. Lacking facts, all we can do is consider hypotheses from eyeball, intuitive observation.

AESTHETICS AND PROFIT

The first hypothesis is that beauty in the environment has a low priority in our economic scheme of things, which stresses competition and minimum costs along with maximum profits. This becomes most evident when considered in the light of two points in the American creed that both private and public propagandists accept with faith; namely, that government is an undesirable actor and that public spending is inefficient. This belief has been propagandized so effectively that most people and most of the mass media regard taxes as a burden, whereas prices are simply prices. Of course, the fact that people must pay taxes but can choose to pay different prices for the things they want enters the equation. Nevertheless, it is a strange society where one seldom hears a good word said about what tax money can buy.

The nation's most critical problems since 1945 have come from ugliness, from ghettos and discrimination, ignorance and poverty, overcrowding and pollution, violence and waste, yet the approach to public spending for curing these conditions still begins with the assumption that real cure is too costly. So we go on treating symptoms rather than causes. It is still much easier for the people who make the decisions to get money for the military and the police, rather than for good schools, more parks, beautiful architecture, the protection of wildlife, or for halting pollution. The thinking is the same as when the nation was struggling to overcome scarcity. The richest people on earth are troubled about the damage that has been done to the nation's soul, and their makers of decisions are still miserly about spending for the things that could revive its soul.

In different words this is what John Kenneth Galbraith was saying in *The Affluent Society*, whose title became a part of the language but whose thesis was lost on many of those in power. The thesis was that the United States had the appearance of being wealthy but not enough of the benefits. We had high consumption, rising production, and lots of material goods. We also had high consumers' debt and we were poor in all those services that we were neglecting because of our emphasis on production and selling. The enjoyment of life was missing. We were in social imbalance; the goods and services produced by the private economy, with its habit of production and advertising to create more production, were far ahead of the goods and services coming from the public economy. It was time to spend more by government.[12]

We are poor indeed in such things as high quality education, good health care, clean streets, parks, productive leisure, attractive public architecture, rapid and safe transport, efficient mail service, planning, of all kinds, and protection against threats in food, pharmaceuticals, water, and air. The wealthiest nation in history could totter dying from poisons to its grave because it refused to start thinking in new ways about beauty and its enjoyment.

The great exception to this generalization is, of course, that the people of wealth and power have frequently spent huge sums on works that either in their own time or later on were accepted as art—the tombs, temples, cathedrals, palaces, sculptures, paintings, gardens, music. Much of this collected art was opened to public access through donation or purchase. Throughout the western world, for small fees, architectural landmarks can be admired along with painting and sculpture in museums; music can be enjoyed. Thus school children are introduced to the great heritage of world art. Such an exception cannot be designated minor. Museums do not make a city, however, and this kind of collected art is only a small part of the aesthetic environment.

WHO SPEAKS FOR THE PEOPLE?

The second hypothesis is that the spokesmen for beauty, all those who accept the "truths" about the aesthetic environment, simply do not reflect the concerns of most of the people who make decisions about how the environment is to be used.

If the custodians of aesthetic quality, the defenders of beauty, really did reflect the interests of most of those who make decisions, then who speaks for the mass of Americans who are far-removed from decision-making in this important area? If they don't speak for anyone except themselves, then they are about as far removed from the real world as an Amazonian Indian. We have to conclude that the artists, critics, architects, landscape designers, planners, and others of like interests do not speak for anyone except people like themselves. Seldom have they accepted the techniques, symbols, and implications of American mass culture.

Our colleague Ralph Huitt, from long study and practice, once said in a lecture that our political system usually has within it, insofar as any issue and decision are concerned, an "aristocracy of the concerned." It is made up of those citizens of any age and influence, inside government and out, who take the trouble to learn about what is going on in their communities, from village to nation, and do something about it. Those who care about the aesthetic environment are, for this subject we are now considering, part of the aristocracy of the concerned. When they band together to exert political influence, they have a chance to win. And it is only when they win that respect for beauty increases and life becomes more enjoyable.

NOTES

1. Fred Logan, "Aesthetic Quality of an Urban Environment, Madison, Wisconsin," "Madison: Aesthetic Quality of Visual Environment," and "A Program of Development of the Aesthetic Quality of Environment in Madison," mimeographed, Institute for Environmental Studies, The University of Wisconsin, Madison, Wisconsin 53706.

2. Philip H. Lewis, Jr., and Associates, *Aesthetic and Cultural Values, Upper Mississippi River Comprehensive Basin Study*, National Park Service, Northeast Region, 1969, Introduction and p. 18.

3. *Ibid.*, p. 20.

4. Paul Shepard traces the history of this respect for nature, pointing out that the garden in Eden was probably the Tigris Valley in the "fertile crescent" where civilization began. *Man In the Landscape* (Alfred A. Knopf, New York, 1967), especially Chapter 3, "The Image of the Garden."

5. Ian McHarg, *Design with Nature* (The Natural History Press, Garden City, N.Y., 1969), pp. 7–17.

6. Madison City Planning Commission, *Downtown: Proposals for Central Madison* (City Planning Department, Madison, Wisconsin, 1970). The author's copy was picked up in the lobby of his bank, a financial institution with a strong sense of one kind of environmental beauty. It has preserved and occupies a historic and beautiful building which is surrounded, except for the State Capitol, by architectural garbage.

7. Lewis Mumford, *The City in History: Its Origins, Its Transformations, and Its Prospects* (Harcourt, Brace & World, New York, 1961), p. 9.

8. John Canaday (captions and commentary by Katherine H. Canaday), *Keys to Art* (Tudor Publishing Co., New York, 1964), p. 37.

9. *Ibid.*, Part IV.

10. Paul D. Spreiregen, *Urban Design: The Architecture of Towns and Cities* (McGraw-Hill, New York, 1965), pp. 1–2.

11. Russell Lynes, *The Tastemakers* (Harper and Brothers, New York, 1949), pp. 21–36.

12. John Kenneth Galbraith, *The Affluent Society* (Houghton Mifflin, Boston, 1958).

chapter

—— 7 ——

SURVIVAL:
WHOLE EARTH PERSPECTIVE

»»»»»«««««

THE REST of this book will be about the social instruments by which man can save the environment—politics and propaganda, business and government, the professions, and bureaucracy. But before we consider these, we must talk of the time during which we contemplate the use of this environment.

We cannot continue to live as we do now. When we dump too much waste onto the land or release it into air and water, we save money momentarily in construction, production, and public services but lose more lives each year from environmental poisons. And we follow a dead-end street. This is a cruel, immoral way to manage human affairs, especially when we look ahead to the time when babies will die in large numbers and the species of man will decline into a few hungry, diseased tribes trying to survive in those few remote places where poisons have not killed.

The span of time that foresight should cover depends on what kind of man we choose to try to save, for man, now with the ability to manage his own evolution, can evolve faster than by natural selection. To speculate about the kind of future man to produce, and who should design him, is too debatable to discuss here. Instead we choose to let nature take its course and to think of a future for us, ourselves, *Homo sapiens*, in all our puny strength but brainy magnificence, high forehead, balanced head, upright posture—the smartest and most superior creature the earth has ever known.

We have been here for perhaps 30,000 years, living at first on the animals we killed and plants that we gathered until some 10,000 years ago we began to plant seed and wait for food to grow. Our cousin, *Cro magnon*, appeared in the Perigord region of France about 25,000 years ago, hunting animals for food, painting pictures and sculpting figures, for reasons unknown. His earlier relatives, *Homo erectus* and *Homo neanderthalensis*, were crouching, slope-headed kin. The present environment should be saved, then, for as long as *Homo sapiens* needs it. When his successor takes over, that next hominid can reopen the question.

There is no easy way to reform the management of the environment. The tasks are too large and too difficult. In order to save the world, literally, the minimum necessary is recycling and monitoring of man-made "waste" and control of population, all handled for the whole earth and all handled during periods of constantly accelerating change.

Anything less than recycling of what we now call waste will always be dangerous. Such products of man's activity, if carefully managed, can be diluted; spread thinly in the soil, air, and water; and kept from being harmful so long as oxidation causes natural recycling. Only two things can happen to the so-called waste that does not recycle from oxidation. One, it will reach a limit beyond which it cannot be stored safely. Two, if this limit is exceeded, disaster will follow. Over the years, man has exceeded the safe limit. He cast a familiar sad account. Now he finds that too much oil has been dumped in the oceans, too many chemicals and dangerous gases into water and air.

Far safer than indiscriminate dumping is recycling. Our colleague John Steinhart once said, "By-products of our society do not vanish without trace. The first law of thermodynamics assures us that they do not. Economists talk about consumption, but in fact we do not consume anything; we merely transform it into a different and less useful form." [1]

The first law of thermodynamics is "conservation of energy." The law means that whenever matter changes form, as by burning, for example, the heat or work produced is proportional to the amount of matter that was changed and no matter is destroyed. It still exists in the same quantity but in different form —smoke and ash, for example. This law is one of the basic laws of physics. All a non-physicist need know is that to a physicist, there are only three things in the universe: matter, energy, and space. The whole physical universe is the interplay of a fixed quantity of matter and energy in space. Therefore, when the first law of thermodynamics refers to the conservation of energy, it also means the conservation of matter, because matter and energy are the same thing in one form or the other.

The concept of waste was born in economics and not in physics, and the way we now identify the matter now called waste proves to be antithetical to life, which depends upon environment. The nature of the earth and universe is physical. Economics is an invention of man.

For life to continue until evolution produces another hominid, man will have to change to an economy in harmony with physical law. What we now call waste is merely a form of matter, and a physical use must be found for it. This creates the urgency for recycling.

Soon, if we are to continue to live, new ways will be found to re-use food, chemicals, gases and other matter now being discharged into water, soil, and air. Examples of a new beginning already can be seen in magazine and newspaper accounts. Some of the same container companies that joined together in 1953 in the Keep America Beautiful campaign to fight litter formed, in 1970, the National Center for Solid Waste Disposal. The center will demonstrate "the technical feasibility and the economics of recycling solid waste back into something useful." A new law allows a federal agency, the Bureau of Solid Waste Management, to guarantee up to 75 percent of financing for a complete plan designed to help a city handle its solid waste, including recycling it. New York State's Environmental Conservation Department plans "giant regional recycling and re-use centers where solid waste from large parts of the state would be brought to a central location, a place that would have around it industries that would manufacture new products from the reclaimed waste." [2]

Dynamics Corporation of America "announces a totally new approach to soil management," a machine that in the process of cultivation allows more moisture to remain deeper in the soil, deposits 75 percent of the crop residue, a solid waste, into the soil, and leaves the remainder blended with surface soil to reduce wind and rain erosion, all in a field ridged for additional protection against erosion.[3]

Boston Edison Company will spend $5 million for a large-scale test of a method for removing sulfur dioxide from smoke stack gases by a method developed by two chemical companies. The sulfur compounds that are to be removed from the stack gases will be shipped to a third chemical company in Rumford, Rhode Island, enough to supply its entire requirement for sulfuric acid. Half the cost of the test will be paid by the federal government's National Air Pollution Control Administration. Some additional financing will come from two utility trade groups.[4]

The Wall Street Journal lately devoted one of its feature stories to recycling. It found that a company formed to collect garbage in Houston and to sell metal and paper waste and compost for re-use was losing money because manufacturers would not buy. For one reason, paper companies had so much invested in forests and pulp-making equipment that they did not want to buy waste paper, although paper is the biggest part of a city's trash. For another, growers of timber get a 10 percent "depletion allowance" when they pay federal income taxes.

Recycling is not new. For economic reasons half of the copper, lead, and iron; 30 percent of the aluminum; and 20 percent of zinc used in the United States is already recycled. Only about 20 percent of waste paper is re-used and less than 10 percent of waste textiles, rubber, and glass. An officer of the National Solid Wastes Management Association thinks the federal government should bring the first pressure for change by buying for its own use only recycled goods.[5]

Owens-Illinois Glass Company agreed to pay 1 cent a pound for used glass in Ann Arbor, Michigan, the first time it has bought scrap in a city where it has no plant. A University of Michigan student group named Environmental Action for Survival, ENACT, set aside two days to collect scrap glass. The company was to analyze how much cost transportation would add to recycling.[6]

Two scientists of the Atomic Energy Commission, pointing out progress in research in ways to harness the energy of the hydrogen bomb, suggest that a nuclear fusion "torch" can be used to vaporize solid waste back into its basic elements. The basic elements, tin, iron, and all the others, could then be used to make more goods. Recycling could be from use to re-use in a closed circle. No by-product would be added. There would be no chemicals, gases, or particulate matter produced in such reduction, and no radioactive waste products, because controlled fusion itself does not produce them.[7]

One of the main costs of operating a chicken factory is for disposing of the manure. For two years now a group at the University of California in Berkeley has been using liquid chicken manure to grow algae. Harvested algae in dry form is a protein rich food that is fed to the chickens that produced the manure in the first place.[8]

The Pure-Way Corporation of Verona and Cottage Grove, Wisconsin, has introduced an enzyme bacterial toilet for human use that produces pure water from human waste. The effluent can be placed directly into streams without violating any Wisconsin regulations. Recycling occurs in a self-contained toilet, without septic tank and drainfield. The expected market would be homes and campsites, indeed any place where an ordinary rural toilet would be too near a lake or stream to suit the law or located on soil that would not be approved for septic fields.[9]

The Dane County (Wisconsin) Regional Planning Commission has before it a proposal to close the sewage treatment plant into which go the waste waters from 70 percent of the county's population. Effluent from this plant now pollutes Badfish Creek, into which it flows. The proposed scheme would involve transporting the sewage to an entirely different place in the county where it would be treated in aerated lagoons then put into a spray irrigation system and used as fertilizer on farms. "These wastes— nitrates and phosphates—are out of place," says John R. Scheaffer, a member of the consulting firm that prepared the proposal. "We've got to shift our thinking of waste as a negative, common enemy, sort of thing to thinking of it as a resource out of place."[10]

And the U.S. Forest Products Laboratory located in Madison on the federal edge of the University of Wisconsin is carrying out research on the city's trash that may lead to a massive recycling of solid waste so that the problem of where to dispose of it will disappear. The laboratory converts wood fiber to paper, including the wood fiber that exists in the form of waste paper. Metal, glass, and other material the Forest Products Laboratory cannot use it sends to the Bureau of Mines and the Bureau of Solid Waste Management, other federal agencies, so they can conduct associated research.[11]

Throughout the nation private groups, some business firms, some governments, and some individuals were either practicing or planning to practice the recycling of waste.

We can find ways to convert the by-products of any human process into something useful. This is the only way we can re-inherit the earth. But to recycle the material we use is still not enough. A system of monitors is also necessary to keep a constant inventory of the earth's resources and to spot any lapse in recycling. When the advanced thinkers talk of the technology for monitoring, they usually dwell on combinations of observation by men and instruments on the ground and instruments in the sky and in space. Infra-red photography has opened up a new method and scope for observation. Observations can be made and reported by telemetry from unmanned posts on land and from buoys at sea, from free balloons, orbiting satellites, and synchronous satellites that remain roughly 23,000 miles above the same place on earth and cover almost an entire hemisphere.

The world's weather is already reported from unmanned satellites. The accuracy of their reports is checked when necessary by an observer on the ground. Techniques for automatically monitoring the animal, vegetable, and mineral resources of the whole earth and for spotting any variance in their conservation are either already at hand or can be developed quickly. The term for such work is "remote sensing;" the instruments, sensors, and a growing field of sensory science and engineering has already been established.

No other subject in the whole earth environment generates as much confusion as population. The confusion is attributable mainly to the viewpoint of the spokesman. If he considers all the possible ways to feed more people on earth and all the empty spaces that can be improved to house them, his tone reassures. If he envisions the damage to nature that overcrowding would cause, he cries alarm. One conclusion is certain regardless of viewpoint. Over a coming span of years, the number of people on this earth must be controlled. If the population grows too large, it upsets all plans, all programs for sane use of the environment. If it becomes too small, there will be too few people to enjoy the virtues that the earth provides.

Population control is essential. Man is the only species that has escaped from the limitations imposed by available food, by crowding, by predators, by changes in the social group, or by man's technology which kills some of the lower creatures. Rats, lemmings, salmon, mice, all the other creatures of the earth, are regulated by natural conditions or by man's technology. Only man can choose.

At the present time, one nation can try to master its own environment, but over the span of time required for evolution, only one-world action can do the job successfully. So many have borrowed Buckminster Fuller's phrase "this space-ship earth" that the words have become trite. There is still truth in the phrase. All life is contained in the capsule of this atmosphere, riding on a whirling sphere in orbit around the sun. What we do to endure is not done for long by nations but by all mankind. Oil pollution of the ocean by a tanker registered in Liberia is harmful to all life on earth, Liberians, Algerians, Frenchmen, and Americans. If the technological revolution has proved one thing, it is that space and time no longer mean very much and that the interests of all men are the same.

Another constant to be added to the equation is acclerating change. Change comes faster and faster. The more we learn, the more we do; the more we do, the more we learn; the more we learn, the more we do. Change builds upon change. Acceleration is the normal pace. Everything gets bigger faster.

Whatever changes man makes to survive on the earth, these changes will have to adjust to future changes. The only unchanging prerequisites to survival are recycling, monitoring, control of population, and world-wide management of the environment. After these the conditions of life can and will change constantly, as we know them to change now.

Life will not be frozen. Diversity will not be lost and new interest stilled. To make life safer will not make it dull.

THE THIRD REVOLUTION

All this, of course, is talk of revolution but of revolution so much greater in its consequence than the change of authority usually meant by the word that there is no comparison. The Third Revolution, to give it a name and capital letters, has the same historical significance as the first revolution, when tools were discovered, or the second, when man changed his life profoundly by growing food. The first revolution, the discovery of tools, became the scientific–technological revolution that reached its peak in the whole network of discoveries, inventions, and techniques that led to the widespread use of machines run by transferred energy, interchangeable parts, assembly lines, load lifters, excavators, automobiles, trucks, airplanes, electronic communication, automation, computers, nuclear energy, and all the rest of what we know and use.

The Third Revolution does not come by violence, nor even by the work of any group. It grows inevitably from the conditions set by the events which produce it, even as earlier changes in man's history came from events such as using tools and planting seeds.

The world is already in the beginning of the Third Revolution. This birth has been observed by the more perceptive wise men. Kenneth E. Boulding opens his analysis, "The twentieth century marks the middle period of a great transition in the state of the human race." He sees our time as the second great transition, the first being the change from pre-civilized to civilized society. The present transition is from civilized to post-civilized society.

Lewis Mumford looks at western civilization since the Age of the Pyramids and sees a Megamachine, the whole organization of power and practice that demands ever more consumption, ever less regard for the humane goodness of life, to keep our present technology satisfied. We are enslaved to this Megamachine even as humans were enslaved in the Age of the useless Pyramids. Now, at last, we realize that the product is not worth the price and we humans are ready to make a change, Mr. Mumford thinks.

John Kenneth Galbraith sees that much of the talk about executive power and the market place has small relation to what really happens in the corporate organization of production and distribution. Technicians, scientists, educators have risen to share the power of decision. The agents of government affect the fortunes of everyone. The change that he describes is so great that it too is part of the Third Revolution.

Charles A. Reich, like Lewis Mumford, finds that the present social system in America satisfies neither justice nor the humane qualities of life. People are beginning to recognize this, he thinks, and will more and more demand change. In what he calls the new consciousness, a new system of ethics will take over to control the results of science and technology.

Henry Ford II, the chairman of the Ford Motor Company, says in a speech:

> It seems clear . . . that neither business in general nor the auto business in particular will survive in its present form. Never before has American business been under such great pressure to change. The real question is whether the changes will be good for the country.[12]

The two most prominent philosophers of the future—they would say of the present—see without pause that all the present trends lead to a oneness for mankind.

Buckminster Fuller, analyzing the effects of technology for years, has seen that increasing speed of travel, wider distribution of electric power, less man muscle required for work, and all the rest are forming one world in which the political ideas of men, such as nationalism, will be changed by the need to make common use of technology's gifts.

Marshall MacLuhan, in his preoccupation with defining the dimensions of man and society in the age of electric communication, thinks that we are returning to the corporate community. In a society of what he calls "resonating information" people live in small groups. Nationalism breaks down. Men soon can prevent harmful innovations.[13]

Unrest, efforts to escape to the country, or into opiates, meditation, group sessions, psychotherapy—into any of the cabinets of transitory peace—can be evidence of the failure of an advanced and dying age to give humans what they need for being human. When proposals are made for something new within the old civilization's frame, we are seeing a desperate effort to keep the dying old system alive when it satisfies no one of feeling. We say the news is always bad, and it is. It reports the failure to resuscitate an old civilization that has had its day. We don't know yet how to let it die without making its final throes even more painful.

The change now underway, uneven as it is, is the great revolution that can provide a new approach to the recycling of matter, the monitoring of resources, control of population, world-wide decisions involving world-wide problems, and adjustment to constant change. Still, it is not certain that man will survive to enjoy the next age. He may not want to be inconvenienced by holding down his ever-increasing use of electric power. He may not want to take the hard way when adjustment is so easy. He may not choose to make the hard decisions to change his social institutions. Man has become accustomed to the idea that he is master of his fate, and he can be lulled into a sense of security so false that it may destroy him, possibly the first of earth's creatures to die of overconfidence.

THE THREAT OF TECHNICAL SKILL

The skill already acquired by man can just as easily be used to survive in a hostile environment on earth as on the moon. This possibility is the first great threat to the preservation of the environment. The easiest way to cope with present trends is to use science and technology to postpone disaster. It is true that no man can foresee what kind of human and what kind of society such adaptation would produce, but we have been this way before and we took the most obvious, easiest way at the time, increasing the danger by adding more automobiles, industries, slums, freeways.

Because the first law of thermodynamics still holds, adjustment to an environment used as a dump is only a temporary measure when seen through time. Sooner or later the poisons will kill all life. So far in our history, however, we have adjusted. A familiar example is the cleansing of dirty water so that whole cities can live for now in health, although still precariously.

Here in brief and unelaborated statement are some of the adaptations man can make *from what he has already learned.* These and others are common in the talk and writing of that strange new profession of futurists who take what can be done now and project it into the future. Futurists deal only with facts as their starting base (in this they are the same as the good science fiction writers), and they deal only with factual possibilities, more likely probabilities, when they look ahead. They form a respectable and fascinating new development in scholarship.[14]

The isolated climate is common in airjet travel, air-conditioned buildings, and air-conditioned automobiles. It became clothing for individuals when airmen began to fly planes at high altitudes without pressurized and heated cabins and reached its present form of sophistication when astronauts left their vehicles in airless space. The whole world of television saw them walking on the moon.

We do not usually think of ordinary combined heating and cooling systems in structures as isolated environments, but they are the beginning of adaptation through technology to a dangerous larger environment. The occupant of an air-conditioned automobile can choose plain outside air, heated outside air, cooled outside air, and if he is in heavy traffic where stinks and carbon monoxide threaten, he can turn to "recirculate" and live within the capsule of his automobile. The future urban air-conditioned building—it has already appeared in Los Angeles—will have air that is filtered as well as heated or cooled and air locks for doors so that none of the foul atmosphere outside need ever be breathed by dwellers safe inside.

The step to individual, radio-equipped, controlled atmosphere suits for street wear would be a short one, technically; but, to quote Neil Armstrong on the moon, one giant step for mankind. Infants who cry into microphones and mothers who respond through plastic head masks will bother the humanists who might have survived to advocate the old communion in nature.

New sources of food will be developed to feed more people. It is commonplace now to talk of harvesting plankton from the seas or to talk of cultivating algae for food. One hears less about growing microorganisms for food or ranching the oceans for mammals. The plant breeders have long been at work to increase the yield per acre from hybrid seed and to increase the protein content of grain. A strictly technological approach to the ominous fact of over-population tends to lessen the threat of starvation, if not the threat to happiness from being alive on earth, and to survival.

But then many things can be done to use more of the earth's surface. Humans have lived on the ocean floor experimentally. A whole sequence of underwater vehicles and dwellings has been designed and tested. Small air-breathing animals have survived underwater without shelters after their lungs were filled with water from which they extracted the oxygen they needed. And men in laboratories and computer rooms quietly calculate how many more humans can be accommodated on earth with the addition of more living space. As for cities above harbors, some architects envision great towering structures built on platforms above shipping lanes and providing all the needs of urban life.

As for the new biology, the futurists are progressively far-out. The first stage is already here. It can be summed up as the practice by some human beings of selecting the fates of other human beings.

A committee decides who will use the artificial organ in a local hospital; not all who need it can be accommodated. A new realm of law has yet to emerge to set community standards for selecting who will live and who will die. In the meantime doctors and hospital administrators ask a few outsiders to help them take the responsibility.

It is not a pleasant task to assign life and death. It can be treacherous when the law is not yet established and suits for damage and malpractice are a constant threat. Committees usually keep very quiet about their work, even their existence. One hears, how reliably none can say, that rough criteria have been established for the choices. A young man of thirty, father of two children, is chosen over a man of sixty-five whose children are grown. A young man of otherwise good health is chosen over a young man, same age, whose kidney trouble is only one of several diseases which indicate short survival. A mother is chosen over a woman with no children. And so on, one hears. It is also rumored that such a selection committee usually includes a physician other than the one in charge of the case, a judge off duty, a minister whose sense of mercy is more trusted than that of other men, and perhaps a nurse or a social worker because they too are trusted to be merciful.

No committee, only two people, the parents, will soon be able to choose the sex of their offspring. The new biology has found that the sperm that determines a male child is different in shape from the sperm that determines a female child. As soon as the technique and tools are found to separate the two, the rest becomes a matter of selective artificial insemination.

Physicians, consultants, "teams," and donors all get involved in choosing which patients will get the next transplants to become available. One of the more poignant comments on the new biology is to listen to television interviews of patients lying in a hospital waiting for the next usable heart. A rare conjunction of two human fates, one of death and the other of hope, must occur to give one person's heart to another person who needs it. Transplanted kidneys, corneas, or blood are easier to come by than hearts. All told, much can be said in favor of the development of artificial parts; the electric heart stimulant, the plastic valves, man-made arteries, veins, and guts turned out by the chemical and metals industries. Less chance is involved in replacing an old organ with a new.

The risk that their children may suffer from genetic defects has long been a nagging fear to many parents. Now there are genetic advisers who can inform them of the chances of mental and physical defects showing up in their babies. Then the parents can choose to take the risk and have children, or be sterilized, or to adopt children. Very soon, it is predicted, instead of combining their own genes, parents will be able to choose from embryos conceived by others and guaranteed within reason to be without defect. The foster mother can bear the embryo in her own womb and give natural birth; the whole experience would be the same as in ordinary mating except that the embryo would be reasonably free of genetic risk. Such medical advances were unheard of ten years ago.

There is also talk of redesigning the human body, by selective breeding, if this makes sense in terms of improving the stock. In other words, humans can be matched for improvement of the breed just as horses, dogs, and cattle are bred by design, and for the development of new breeds. Also under investigation is cloning, a procedure for duplicating individuals, cell for cell, which has already worked in smaller species. The nucleus of a cell of the individual to be reproduced is placed in an unfertilized egg of a foster mother. The resulting offspring is a total duplicate of the individual from which the cell nucleus was taken. As many duplicates as wanted can be produced.

Another distant expectation involves integrating living tissues with machines to make the machines more competent. A few futurists talk of integrating a human brain with a massive computer to produce a machine that could do more than solve mathematical problems quickly and type rapidly. Thus some future generations of computers (the term "generation" is used to designate a stage in computer design and competence) would become living things, at least in part.

Most familiar to laymen in all this futuristic talk is a machine that attaches to the human body so that the living actions of the one are translated into mechanical actions by the other. Best known is an artificial limb that is made to work by the very nerves and muscles of the wearer. These machines lie as far ahead as man's needs and patience extend. They can carry heavy loads over rough terrain. One such machine, already built, consists of hinged legs and feet that, with a man attached, will walk wherever a man can walk. Others can handle dangerous materials at a distance and behind protective glass or do outside work in the ocean deep as directed by a man inside his insulated vehicle.

Man's technical skill may be the very danger he should fear most. He can build a machine or find a biological technique for nearly anything he can imagine. The more he learns the more he does, the more he does the more he learns, the more he learns the more he does, the more he does the more he learns; so change accelerates and tomorrow comes sooner until tomorrow is always already here. But meanwhile what will have happened to the established values of yesterday? Who can say? Except that they will have changed.

Life can be made safe and healthful for a time, and men can be as solidly content as the most superior herd of Angus cattle, all the meat distributed just right, all the bones and nerves at prime, the diet mixed finely and medical attendance prompt with spare parts. Somehow such men will be a different breed from the imperfect specimens who brought us this far. They will not even know what choice they had back in the late twentieth century. And they will still be on that dead end street of exhausting the earth's resources.

THE PROMISE OF TECHNICAL SKILL

Technical skill can be used as well to recover and keep the kind of world that will allow humane values to live. The skill of science truly is not good or bad. Its use is. Heat can be used for healing or killing, a point so often made, and it has been used both ways.

In the watchful night when two men first walked on the moon, more than one American said, "If we can do that, why can't we clean up the slums and get rid of pollution?" It was—and is—a good question. The answer is so simple that it inspires argument. Technically, we can! The machines and techniques, the trained men, the method of management that put men on the moon can be used to clean up the environment. These are technical skills. They are not used on the environment because Americans and other men do not choose to use them.

Americans, or more particularly their leaders, were willing to use technical skills to land on the moon because such an enterprise did not threaten any of the accepted social institutions. Instead, it was good for profits and wages, good for the Gross National Product, good for the public and private bureaucrats who had a share in the contracts and activities. It was good for national prestige in the propaganda competition with the Soviet Union. It was a welcome distraction from prolonged, unwise, and unpopular involvement in other nations' wars. It was good for national morale at a time when Americans were discouraged by unsolved problems of pollution, civil rights, cities, the future, the continued cold and hot wars, all in a time when a generation had a right to expect serenity. Neil Armstrong's step was a healthy reviving change for Americans, who could be proud once more of an achievement that did not call for killing.

A giant leap is technically possible in managing the environment. The barriers to cures and preventions in the environment are seldom in the lack of technical skill. If the skill is not yet available, it will be shortly. One of the conspicuous traits of a sci-entific-technological age is that once a technique is needed it will be discovered. When well-informed men discuss the need for re-form, they almost never mention techniques alone. They also talk of social practices. Here is the second great threat to the re-tention of life on earth.

THE SOCIAL THREAT

Our present social institutions and practices probably cannot handle the enormous task of total recycling, total monitoring, and total control of population on an earth-wide scale during constantly accelerating change. Some examples of reasons for doubt stand out at once. Nationalism is one. The tendency of de-veloping nations to follow the same historical cycle as the devel-oped nations, making all the same mistakes, is another. These are world-wide threats.

For the United States the questions are similar. Americans, faced with the rape of their environment, worry about whether the nation, states, counties, cities, townships, and special districts, can ever come up with solutions when they spend so much time trying to agree among themselves; or, indeed, whether govern-ment regulation over private polluters is adequate to stop pollu-tion; or whether the distorted distribution of taxes, in which the national government gets most of the revenue to spend mainly for military use, will ever allow us to clean up the environment.

The present uneasiness about competence and direction in the

United States is traceable to indecisions about some very basic social philosophy and practice. In this small space these indecisions
can only be listed in outline. Brevity should not demean their
significance.

1. We are caught between centralization and decentralization,
and we make compromises between them but no firm decisions
or definitions. In government, the states require deference from
the central, or national, government because of the states-rights
heritage that has existed since the time the Constitution was
written and approved. And within the states, the counties demand deference. Oddly, for an urbanized society, the cities are
the stepchildren, begging for crumbs from the county, state, and
national authorities but getting small sympathy and little response.

Now we cannot say unequivocally that centralization is more
efficient than decentralization. In practice the sensible answer
usually depends on what is being decided. Some matters of the
environment, for example the appearance of junkyards, are local;
others are regional, the condition of a lake, for example; others,
such as automobile pollution, are national; and still others are
international—radioactive fallout, for example.

No easy generalization can be made about where all decisions
should be made. In the same moment, one generalization can be
offered: When indecision about the location of authority causes
incessant negotiation and confusion among local, state, and national agencies, the work of society is slowed and often natural-
ized by the necessity to compromise.

2. We have not decided the proper balance between democracy and direction. In a nation where work in committees begins
in the third grade, it follows that many citizens believe that all
decisions should be made in a democratic manner, with each person who has an interest respected for his view. Experienced realists know that few committee decisions are as firmly based on evidence as they are the results of compromise: yet government,
business, education, and private groups spend much time and
make many decisions in committees.

Two consequences of this indecision about democracy and
direction are serious. One, many of the social decisions that are

most seriously needed—planning, for instance—are not made because in the process of democratic behavior no one suggests them, and if suggested, no one can win the argument for them. Two, in the delusion that democracy is truly practiced, and that the American system is really as simple as it was taught in the ninth-grade civics classes, too many Americans fail to watch their public and private officials and hold them responsible. By the time a citizen discovers that a decision was made without his knowing about it, he may be fighting a war or suffering from pollution that he would have tried to prevent had he had sufficient information in the beginning. Nothing so cripples the American citizen as his belief that as citizen and consumer he really has a voice and can control the decisions that affect him. He believes that he lives in a democratic state, whereas he really lives in a bureaucratic and pluralistic state that includes many public and private agents and groups.

Another way of saying this is that when any society begins to live by sacred truths instead of reality, it becomes subject to the will of those who know that sacred truths are myths and who know how to use them to get what they want: shamans or bishops, the "royals" or elected executives, legislators or judges, public relations men and advertisers, bureaucrats, generals, or scientists.

3. We have difficulty deciding when emergencies are over. Time and again we adopt measures to meet an emergency but we never discard them.

For example, the atomic bomb was regarded, in 1945, as a temporary weapon. The United States dropped two bombs to hasten the end of the war with Japan and to lessen American casualties if we had to invade. True, we tried to get international control of atomic weapons and failed. But we also continued to develop, manufacture, and store nuclear weapons until we reached a new wonderworld of over-kill. The early argument for post-war bombs was that we had to be prepared for any new emergency.

After the two big powers had both exceeded over-kill many times over, and were fearful of challenging the other, we found a new reason to continue to produce nuclear weapons. We had to have more and newer ones in order to preserve our bargaining strength at conferences with Russia over limiting the use of nuclear weapons! This new reason will become permanent unless we change our ways.

This tendency not to end emergencies becomes most serious when the emergency is war, for war drains the public treasury until only small change is left for the environment. In the curve of expenditures by the federal government, the peaks reached during wars are always followed by plateaus that are higher than they were before. In recent time, troops have been involved after the fighting stopped. It was true in Western Europe and Japan after 1945 and in Korea after 1953. Such is the prospect too in South Viet Nam after American forces stop fighting there. Troops overseas in large numbers mean larger expenditures than would be needed for a home guard.

4. The deeper reason for spending on war is a separate indecision. We have not decided, as mankind, whether we are to emphasize force or more humane purposes. So long as America and other nations subscribe to military action in the ancient way, but with always newer weapons, we will sacrifice some of the good life. It costs more and more to kill the people of other nations.

5. In all our institutions, including government, we react rather than anticipate. This indecision has been argued since the nation was born. The proponents of planning, or anticipation, say that government can know what lies ahead with enough certainty to make those moves that will prepare for it. Alexander Hamilton thought so in his *Report on Manufactures*. The other authors of the *Federalist* papers agreed. Opponents argue that the future holds events that cannot be anticipated and that a nation does better not to plan but to be flexible in reacting to events.

The harm comes from doing neither very well as a result of indecision. Insofar as care of the environment is concerned, we lose time and increase the damage each time we wait to discover new forms of pollution and then waste more time before we react.

6. We are undecided about how to make decisions and do not yet have firm answers about the best methods. Very briefly, among those who study this question, there are two broad tech-

niques involved in making decisions. One has been mentioned in part as the pluralistic, or democratic, method in which competition among participants is expected. The other can be called the executive method. When practiced, which is seldom, an executive with authority gets all the evidence and advice he wants and then makes the decision.

In either method, a decision can be incremental or innovative. The incremental decision comes from a history of facts and events, and the latest decision builds on the record. An innovative decision is reached without regard for the past.

Some think that an incremental decision is the only kind that man with a history can make. They say that man cannot ignore the history that made him. The point is important in any decision concerning environment. If we accept the accumulated past as an important part of the decision, the plan of action proceeds from what has been and not necessarily from what should be. An innovative decision can produce action that starts without restraints from the past. It will cost more, of course, and it may prove to be wrong in the future, but some other people think that only by innovative thinking can the "good" be obtained.

Another subheading under making decisions is "systems" thinking as opposed to thinking based on segments, preferences, and intuition. The systems approach means that a whole is seen instead of a part of a subject. Within the limits of human ability, all parts in a whole are related to one another. The systems approach is advocated by engineers, administrators, and scientists, who know the advantages of using computers to accommodate many different variables, as they call the parts to be related.

7. The last indecision is the most serious of all for the future of environment. We have not decided whether we will continue to accede to ever-increasing production and consumption or whether we will choose the less threatening but greatly different system of humane standards. Because the present system of technology can always produce more, we decided that it should. The price we have paid is damage to the environment, loss of the benefits of handcraft, and the indignity of high pressure selling aimed at making sure that the new production be consumed so that production can continue to increase, and so on. This basic

question of the future is the final concern of Mr. Mumford, It is the uneasy, undefined worry of all Americans who ask such questions as whether we really need to double the production of electric power every ten years and constantly produce more automobiles, highways, and television commercials.

All these indecisions require debate and analysis. Books could be written about any one. They are listed here only to show what is really at the heart of whether we will be here until the next evolutionary change. Our assurance of a future on this small planet depends on our undoing such indecisions by making the right decisions.

THE HANDICAP OF SIN

As so many others through centuries before him, Tristram Coffin says in a letter to the magazine of the United Nations Association, in a gloomy search for secular salvation:

> What is the course? I frankly don't know. None of the forms of government or political philosophy men have devised work out too well, and I think this is because of our obsession with form and shape. We think, ah, if we could just tinker with the form, overthrow the czar, install a President, hold democratic elections, things will be OK. The problem is not the form, but man himself.[15]

The understanding of sin was easier when theology provided the answer. Man had inherited original sin from Adam and Eve who disobeyed their Creator and ate the forbidden fruit. The Adam story had to fade with the discovery of a line of hominids of which man is the only specimen still living. The trouble is that while man has lost the theological simplicity of it, he has not stopped the practice of sin.

"An entire generation of thinkers and writers," said Charles Frankel in 1955, "has rediscovered the truth in the ancient doctrine of original sin." He was reflecting upon a recent time, upon "a spectacle of human wickedness raised to a new level of sophistication and efficiency." Death camps and wars, pollutions, and punishment of the helpless still have to be explained by motives that lie outside the notion of man as a kindly and civilized animal. Mr. Frankel turned to a modern theologian, the late Reinhold Niebuhr, to try to find an acceptable definition.

> Thus, there is in man, as Mr. Niebuhr sees him, a connatural perversity, a discordance in the human spirit that echoes a larger disharmony in the relation of man to the universe. And this produces a basic ambiguity and irony in all human history. For the source of all man's achievements is also the source of all his wickedness and folly. His vices and his virtues have the same origins. And so there is a taint in whatever humans do. . . . Human wickedness is not the product of bad education, or mental disease, or unjust social conditions. Specific human misdeeds may be, but not the fact of human evil. This is *original:* it is uneliminable from human nature in all times and circumstances, and is the ultimate source of bad education or mental disease or unjust social conditions.[16]

The statement is not scientific, perhaps not even provable, but science has given us no better definition of the ageless and repeated observation that enough individuals in any human society will behave with evil toward their fellow men to make sin a constant trait. Plain meanness, selfishness, neglect of others has to be counted as another handicap in saving a good environment.

Sin leads a manager to dump pollutants into the air and water at night when he cannot be detected. It leads a public prosecutor to overlook pollution by an industry that employs voters who elect him. The eyesores left by citizens after picnics are in their degree a reflection of sin, and until people are ready to police themselves for the sake of others, even in small aesthetic pleasures, there may be small hope of true conservation of the environment.

THE IMMEDIATE FUTURE

There is no way to instant salvation, and we should not expect to get there except by knowledge, persuasion, and considerable fear and trembling. All men everywhere on earth misuse the environment to some extent. The most advanced men in the most advanced industrial nations are only now, after 25,000 years of learning, becoming aware of the swift danger of too much misuse. Yet only some of them are ready to talk about ways to reform their practices and to help with such high priorities as birth control. No one can say whether the advanced men of the earth can change their own practices and morality in time enough to save their lives. And no one can say whether these most advanced men can inform and persuade the less advanced not to make the same mistakes they have made.

Uncertainty about survival can never be removed for a species. The dinosaur could have expected extinction if it had been educated. As a thinker, man can expect at some time to disappear as now formed and to evolve into some other type of hominid. He may change suddenly. Some cosmic bump in the night, another change of hemisphere by the magnetic pole, a violent insult by new viruses, many events can cause a fast evolution of man. Uncertainty is normal. A certain future is not the goal.

Yet something must be done and fast. We began action to survive in a hostile environment some two hundred years ago. We began to discover vaccination, bacteria, and how to purify water. We are not unused to the idea that survival is important and that physicians will use all the life-saving drugs and techniques that science-technology gives them. Under threat of death from diseases that could be prevented, men accepted such new practices and moralities as compulsory vaccination, surgery, fluoridation of a city's water, and the mass inoculation without permission of all men and women in the armed forces.

The next two hundred years, rather the next fifty or a hundred years because of accelerated change, can see just as great a reform of attitudes and practices concerning the environment. The change has already started in actions by government, some businesses, some universities, the media of communication, and the young, who become in a startlingly short time the next responsible generation. When the young agree that something must be done, something has a good chance to get done, or at least to get started, within the next fifteen years.

The immediate future work of all good men is, therefore, to do faster more of what has already been started and to make sure that the young are encouraged to make their own changes later. More attention to all the pollution, more cleaning up and keeping clean, more research into new ways of production, more teaching and persuasion to conserve, and all the other actions that we know to be good, make up the work to be done in the immediate future. A lot less sin would be a great help.

SOME VIRTUES OF THE AMERICAN WAY

As a nation we Americans have been so foolish in foreign involvement which cost dearly in money and grace that we tend to forget our virtues. Almost paranoid, and deserving to be, we see other people scorn us, insult us, nag at us for being silly about foreign policy, and we agree. The blot of foreign policy becomes a function of shame, a harelip that makes ugly the personality of a girl who would otherwise be beautiful. We are so much brighter about so many other things that our incompetence in foreign affairs is a self-induced disability.

We are used to saying that we have the most advanced physical technology ever known. We seldom recognize that we also have what might be called the most advanced social technology ever known. We produce more data about ourselves and our surroundings in bigger computers than any other people. We have a system of easy credit and easy banking that confounds other people. Gripe as we like, the social technology that can plan, finance, build, and maintain the highspeed highways and the nets of city streets that carry our millions of automobiles with increasing safety is a remarkable series of inventions.

A French journalist, not an American, told us that the social inventions of American business had made it not only superior to European business but an outright challenge to take over Europe's economy. We Americans could think and manage over large geographical areas, he said; we could ignore national boundaries. We had worked an alliance of business, government, and universities that brought benefits to all. Americans had taken the controlling lead in Europe in the newer industries of computers and integrated circuits. And toward American business in Europe, not even the grieving young could lament that it was another part of a military-industrial-educational conspiracy to maintain war. We had no help from the armed forces of Europe, which were smaller anyway and less expensive than our own at home—and in Europe—and therefore had less power. Most impressive for a nation that expects always to pick up the check, most of the capital for American business in Europe came from European sources.[17]

The social invention that produces America's mass communication of information and opinion, proposal and criticism, is amazing. It exposes 200 million Americans to news that is quick, as accurate as professional reporters can make it, and usually labeled as opinion if it is. This news is uncensored except in the quiet deletions made at the source. It comes by a wondrous social organization of print, radio, and television, once or twice a day by print, once on the average every hour by radio, twice a day by television, and continuously by television when the nation is in agony or suspense.

Such accomplishments as these have given Americans a blend of national attitudes that makes us competent in all work save foreign affairs, and our error here may be that we inherited the task from a different world and time and proceeded to handle it with all the boldness and skill of a people who thought that anything could be done by military and economic weight and by machinery. We were wrong.

Our better side is the important one for saving the environment. We expect change as normal; we do not fight it. We know that we can do anything we want to do except win guerilla wars in situations that we do not understand.

We are an open society; no man or class is in charge. We listen to anyone who has a good idea. The politician, the workingman, the banker, the industrialist, the farmer, and the woman in slacks all put their pants on the same way. It is just as important to save a poor man's life as to save a rich man's life. The only special people are children and the young through high school. They may have nearly anything they want, and most of them grow up mature in spite of their early favors. Since the American Depression of the 1930's the idea of social conscience has been accepted in America. We like to hear politicians and businessmen, preachers and professors, talk about duties to the community.

In all, we Americans have the basis for more psychological security than any other people on earth. We need to understand this and to regain national self-confidence. For Americans have to lead all the other people of the earth in saving the environment.

In this stage of the Third Revolution, while national boundaries are still firm, we can do it by example. In the whole-earth phase to come, we can do it, not as Americans, but as people who learned the most about technology and who, because we have self-confidence, can have the courage to be kind.

NOTES

1. John Steinhart and Marti Mueller, "Search for the Future," unpublished report for the Ford Foundation, Feb. 1970.
2. *The New York Times,* Nov. 15, 1970.
3. *The Wall Street Journal,* July 31, 1970, advertisement.
4. *The Wall Street Journal,* July 7, 1970.
5. *The Wall Street Journal,* June 23, 1970.
6. *Wisconsin State Journal,* May 31, 1970.
7. *Wisconsin State Journal,* Jan. 22, 1970.
8. *Scientific American,* November 1970, "Science and the Citizen."
9. *Wisconsin State Journal,* August 16, 1970.
10. *The Capital Times,* Madison, Wisconsin, Nov. 20, 1970.
11. *Wisconsin State Journal,* March 22, 1970.
12. Kenneth E. Boulding, *The Meaning of the Twentieth Century: The Great Transition* (Harper & Row, New York, 1965); Lewis Mumford, *The Pentagon of Power* (Harcourt Brace Jovanovich, New York, 1970), also "The Megamachine," *The New Yorker,* Oct. 10, 17, 24, and 31, 1970; John Kenneth Galbraith, *The New Industrial State* (Houghton Mifflin, Boston, 1967); Charles A. Reich, *The Greening of America, How the Youth Revolution is Trying to Make America Livable* (Random House, New York, 1970), also "The Greening of America," *The New Yorker,* Sept. 26, 1970; Henry Ford II quoted in *The Wall Street Journal,* Nov. 16, 1970.
13. Both Mr. Fuller and Mr. MacLuhan flow with ideas and are difficult to footnote precisely. This author heard Mr. Fuller's ideas in helpful conversations from 1942 until early 1945. His books, especially *Ideas and Integrities: A Spontaneous Autobiographical Disclosure* (Prentice-Hall, Englewood Cliffs, N.J., 1963) and *Utopia or Oblivion: The Prospects for Humanity* (Bantam, New York, 1969) include much of his philosophy. I heard Mr. MacLuhan's point about the decline of nationalism in a television interview in 1971.

It can be derived from *Understanding Media: The Extensions of Man* (McGraw-Hill, New York, 1964), Chapter 18, on print and nationalism. It is explicit in MacLuhan and Quentin Fiore, *The Medium Is the Massage* (Bantam Books, New York, 1967), p. 16:

> Electric circuitry . . . pours upon us instantly and continuously the concerns of all other men . . . Its message is Total Change, ending psychic, social, economic, and political parochialism. The old . . . national groupings have become unworkable.

14. Three books are a good sample of futurist literature: Herman Kahn and Anthony J. Wiener, *The Year 2000, A Framework for Speculation on the Next Thirty-three Years* (The Macmillan Co., New York, 1967), John McHale, *The Future of the Future* (George Braziller, New York, 1969), and Alvin Toffler, *Future Shock* (Random House, New York, 1970).

15. *Vista*, Sept.–Oct. 1970.

16. Charles Frankel, *The Case for Modern Man* (Harper and Brothers, New York, 1955), pp. 85, 89–90.

17. J. J. Servan-Schreiber, *The American Challenge* (Antheneum, New York, 1968).

part 2

*The Institutions
of Social Change:
How They Work,
How to Use Them*

chapter

——— 8 ———

POLITICS AND PROPAGANDA

»»»»»»«««««

ONE OF THE FIRST lessons to be learned by the future leaders of America is that relatively few people manage the affairs of society. The few are those who emerge from the crowd at a certain time for a certain decision and who work for their cause. Sometimes they hold titles. Some of them are public officials, some private citizens. Their incomes vary. Status has less influence among them than interest and ability.

Keepers of the myth can be assured that no threat to the American way will come from the successful few who manage affairs. They like freedom and the idea of rotation of officials; they like the flexibility given the Constitution by judicial review; they like a free press and free enterprise. They like competition. Having learned to succeed within the present system, and knowing that no other system in man's history has worked as well for as many people, they are not about to change the system. Threats come from those who have not been able to get or to keep what they want. Most significant, the successful people learn the two great truths of society. First, change is constant. Second, deliberate change is made to happen through politics and propaganda. They learn the techniques of politics and propaganda in order to influence change.

PERSUASION

Politics and propaganda are never separable. They both aim at persuasion—to persuade voters in an election, to persuade mem-

bers of a legislature, to persuade a board of directors, to persuade
the necessary number of executives or union officials or members
of a faculty. Politics occur in any conceivable social group and
are not limited to government. The politics of a university are
just as real as the politics of Congress, the politics of a church
congregation as real as the politics of city hall. Propaganda is any
effort to persuade others to a chosen opinion by presenting them
selected facts and ideas. It can be distributed in the mass media
or by word of mouth, by national network or by face-to-face talk.
Once the few who manage the affairs of society and who know
the techniques of politics and propaganda agree on the need to
save the environment, it can be saved. This is one great advantage in a polity such as the United States. The successful are
much alike wherever they are.

When a majority of them begin to pay attention to the same
thing, they will all reach much the same conclusions about what
should be done. No blind dogmas will handicap them.

Once the attentive few have recognized that something should
be done, the arguments will be more about means than goals.
The solutions will be found in competition and compromise but
mostly through agreement among a majority of the participants
and acquiescence among the rest. Americans in this way have
made some great changes from the past. Their leaders wrote the
Declaration and Constitution, accepted judicial review, held the
union intact, created new states, built our present social and
physical technology. The few who manage the American system
can save the environment in this one nation and set an example
for the entire earth.

CONSENT OF THE GOVERNED

Only one constraint surrounds them. They must do whatever
they do within the tolerance of the mass culture. In America this
means that they must act in moderation, walking down the middle way, declining to be associated with extremism. Since the
techniques of survey research were perfected in the 1930's, political scientists have had recurring proof of the influence and appeal of the middle way among the majority of Americans.[1]

The decision-makers know this too. "The people" must consent for any new measure to be effective. In a free society, with a free press, the limits of possible solution are truly set by the majority of the members of that society. Only rarely does the majority take direct action, not even in the election of a President, except when the winner gets a majority of the vote cast. Almost never does the majority take the initiative. It is not organized to take the initiative. The majority gives consent to the action of its leaders. It abides by decisions. American leaders and "the people" have shown that they are concerned in general about the subject of environment. How much they will support a particular remedy depends on whether that particular remedy has been recognized and accepted by a sufficient number of people who are strategically placed to get attention.

A SUBTLE AND ELUSIVE THING: TIMING

First, an issue must have come into its time. Ignorance of this fact probably causes more frustration and more disappointment with American politics than any other part of the process. A citizen who wants a change that he thinks important, but a change that most of his fellows do not think important, is unhappy. He blames the system. What makes an issue timely? One of the mysteries of political science is that for such a practical question we have very few firm answers.

Casual observation reveals that leaders—known as opinion-leaders—start the process. Editors and reporters see a new significance in the issue and start giving it space. Business executives start talking, along with teachers, preachers, farmers, workingmen, the girls in an office, a man at the open table in a downtown club, a woman in a car pool on the way to work, a high school student in a class discussion, or the President of the nation.

Anyone can be an opinion leader if he states an opinion. For his opinion to spread, it must be taken up and expressed by other opinion leaders. When enough of these leaders agree that an issue is important, and the rest of the members of a free society acquiesce in this opinion although they do nothing to promote it, the issue becomes important.

This much descriptive theory is probably sound. It tells nothing, however, of why interest in an issue spreads under what circumstances. For nearly any issue that becomes prominent one can find that earlier leaders had tried to raise it and had failed. Consumer advocates had worked in Washington for at least thirty years before Ralph Nader, saying many of the same things that Nader said successfully later. They received little attention. If by hard work they were invited to testify before a congressional committee, they made no news. For years the Food and Drug Administration had been condemning batches of food, some produced accidentally by the most eminent companies, for hideous reasons such as the presence in selected samples of rat hair or mouse dung, and the mass media paid no attention. Now such exposures are news, meaning that now the opinion leaders think that the mass audience is concerned about the quality of what they eat and the responsibilities of food processors. For at least a hundred years, warnings have been issued about the destruction of earth's environment. The facts were plain to anyone who looked, but the warnings were not news. Only in the 1960's were enough people interested to make the issue news.

Having said that timing is important to catch an issue in its tide, what can one say in general about the recognition of a tide? Not much. Opinion can be measured for breadth and intensity by the techniques of survey research, but unless the issue has become known to most of the people questioned, "don't know's" help only to indicate ignorance and lack of interest.

The late V. O. Key, Jr., master scholar of American politics and public opinion, discerned different kinds of consensus when in 1961 he analyzed data from survey research, although he found supportive data "not copious," having said earlier that "to speak with precision of public opinion is a task not unlike coming to grips with the Holy Ghost."

Supportive consensus upholds policy already adopted and actions taken under it. Thus social security had consensual support from its beginning, and public officials, especially the President, usually have support for their decisions. Permissive consensus allows leaders to adopt new policies. Sometimes it is formed long before the action is taken. A consensus existed in the public to admit Hawaii as a state for fifteen years before Congress acted.

Yet the issue was not strong enough to make Congress act. A consensus of decision, as Key calls it, is in the classic definition of democratic debate and discussion. In the months before Pearl Harbor such deliberation occurred and the majority of opinion swung from decidedly against American involvement to favor of help to England. There was still a division, however, which the Japanese erased by their attack at Pearl Harbor. These are all degrees of public acquiescence or consent.[2]

In practical terms we can say that leaders in the movement to save the environment already have the question in the main political stream. Abruptly, as the tides of issues run, public awareness of the environmental crisis rose in a few years. Some day perhaps the historians will be able to trace what happened. At the moment this is not clear, except to say that Rachel Carson's *Silent Spring* will be one of the landmarks as will such shocking events as the wreck of the *Torrey Canyon*, the Santa Barbara oil leak, and the public debate about the S.S.T.

A worried awareness of a threat is the beginning of cure. Awareness makes it easier for leaders of the politics of cure to find the consensus to support them.

THE BASIC ISSUES OF ENVIRONMENT

It appears now that the particular questions of environment, such matters as the various pollutions on which we are working as well as the ultimate physics of how much waste can be oxidized safely and how much access we keep to the sun, are subordinate to four issues that recur in discussions of all the particular questions. The main issues in the politics of cure are:

1. How much time do we have to change the practices that now threaten to destroy the environment? It makes a difference whether we think and act about the environment in terms of years or decades. And once the trend is turned, it may take decades to reach safety. We cannot permit the trend to decline any longer.

2. Shall we continue to gamble on the future? Until now our practice has been to adopt any new technology that promised more speed or profit or military power or employment. Later, if the technology proved to be dangerous to the environment, we

began to talk about regulating it. Another approach is to make the best assessment possible of the future consequences and control the introduction of new technology.

3. Who will pay for the cures? We will say more about money in the next chapter. Here we need only list the question. Because Americans make such a distinct, though wrong, division between public and private money, this issue will continue to affect the movement to save the environment.

4. How will the cost of the cure and protection of the environment relate to other costs? This question becomes most troublesome when the issue of environment collides with the issue of war. War is already the largest consumer of money. Restoring the environment may well become the second largest. If we have to decide to reduce spending for war in order to spend more to save our lives, the issue then will be loud and clear. Decision after decision will allocate money to war (and of course the preparation for war or what the government propagandists carefully call national security), to environment, and to such other standing causes as health, the poor, highways, education, civil order, justice, and never will there be enough to give to each all that its advocates think they need.

For these and for all the subordinate issues, the people who manage the affairs of society will find the tides of consensus by constantly listening and asking questions. They will use consensus to seek change through the use of persuasion, or in other words, through politics and propaganda.

PEOPLE

People are at once the source of awareness and the objects of propaganda. They develop the facts of politics known to political scientists by such words as consensus, expectations, demands, support, and intensity, stability, or latency of opinion. People muddy the affairs of society with the consequences of sin. Men and women, not squirrels, toss out the beer cans, throw away the cigarette packs, find it amusing to throw an empty bottle into a delicately colored, rare, warm spring in Yellowstone National Park. People try to evade the cost of cure and commit the crime of pollution.

People also bring virtue into social decisions. Public and private executives who manage their affairs with the environment in mind are virtuous. So are members of legislatures and all others who take the time to try to save the environment.

This much is clear from living in society. It is not so easy to understand people as the objects of propaganda, for then they become complex psychological networks quite unlike the pastoral ideal men from a pre-Freudian past.

"New theories of psychology," says Key, "brought new conceptions of the nature of man, conceptions that made him a nonrational creature of subconscious urges and external suggestion." Certainly those of us who went through courses with Harold D. Lasswell at the University of Chicago in the 1930's, and learned the melody of what he said even though we did not learn all the words, could never again think of a voter or a consumer, or his leaders and bureaucrats, as the clear-headed rational citizens postulated by Thomas Jefferson.[3] Lasswell's citizen was the new natural man. He reacted to ideas, people, problems, events as much with subconscious emotions as with a rational mind, if indeed a rational mind were possible. This citizen was discovered in the study of Sigmund Freud. He craved safety, income, and deference and felt deprived when he was threatened. He reacted from a bundle of motives that had been created by his total experience from birth. Those citizens who sought positions of authority were moved by the same collections of emotion based on experience. They needed power to compensate against deprivation.

Most of Lasswell's hypotheses from the 1930's were proved in later research. He took the pragmatic observations of such men as John Dewey and Walter Lippmann (in his role of social philosopher) and gave psychological explanations for them. They knew that citizens did not behave as they were supposed to and that the public was what Lippmann called it in a title, *The Phantom Public*. Lasswell said that by the nature of man, the rational citizen who reacted to evidence only by weight and by virtue could never exist because no such man was possible. People really do think with the "pictures in their heads" that Lippmann talked of in *Public Opinion* (1922), still one of the best books ever written on the subject. They cannot perceive all the world around them but choose only those aspects for which they have already conceived a picture. They live by symbols and slogans more than by rationality.

If Lasswell opened the door to post-Freudian psychology in political science, Murray Edelman added the idea that not people alone but all the institutions and processes of government are symbolic. While conventional studies emphasized how people get the things they want through government, he analyzed the way "politics influences what they want, what they fear, what they regard as possible, and even who they are." [4]

Voting is more a rite that allows voters to feel involved than it is an influence on policy. All times are times that try men's souls, and whenever leaders want or have to do something they think may rouse resentment and resistance, they say that the crisis of the time requires it. The incumbents in public office are regarded as leaders simply because they are in office. Public officials do not have to succeed to get support; they must only try to do something. Leaders have to identify with the approved community symbols. But leadership is defined always in terms of a situation; not by what the leader does but by the way followers respond. If they follow, leadership has been exerted.

The citizen sees the public official not as he is but as he appears to be from the symbols with which he is identified, and because national officials are the most prominent, the citizen will "know" national officials better than he "knows" local officials. The police power can be more trouble than minor violations of the law when it is enforced strictly by the rule. Policemen expect speed laws to be slightly violated by everyone, and they are. The notion that government's regulation of private business is a meeting of adversaries is fiction; if government did not regulate, business would have to regulate itself to get the same order and safety for itself that government now provides.

Such radical conclusions, and many more, Mr. Edelman makes quietly, citing evidence for their truth from research in psychology, anthropology, and political science.

PROPAGANDA

The principal technique of politics is propaganda, the use of communication to persuade.

Harold Lasswell and associates once happily invented a sentence that includes all the subheadings of propaganda with which professional analysts and conscious practitioners deal.[5] It asks: *Who* is saying *what, how,* to *whom,* with what *effect?* "Who" is the *source* of propaganda, "what" the *content,* "how" the *media* of transmission, "whom" the *audience,* and "effect" the *change* or *no change* in opinion in the audience.

Many Americans still think that propaganda is a dirty trick that has no place in politics. It is a common practice for one propagandist to accuse an opponent of using propaganda, implying that this tactic is unfair and misleading. Realists are aware that nearly every communication they receive by any means can be and usually is propaganda. One big exception is art for art's sake, but not art as propaganda. In the popular arts many lyrics of popular songs are propaganda against war, for brotherhood, for and against the use of drugs. Another partial exception is education, when the purpose is to transmit skills, but not when teachers are delivering arguments for good manners, hygiene, safe streets, and the benefits of the American way.

Propaganda is any message for which the content has been chosen for the purpose of persuading an audience to think or do what the source wants them to think or do. It is used to get people to a church supper as well as to incite a mob to burn a ghetto. The technique is not good or bad. Its use may be.

To save the environment through politics and propaganda, advocates must know how to use propaganda effectively. First, they must know the elements of propaganda and some of the elementary facts about how to use it. Moreover, they must know how to detect and analyze propaganda as well as how to use it. There is nothing mysterious about the subject. Some experience and some awareness will make any person able both to use propaganda and to be immune to it. The best approach is through the sentence from Lasswell and his associates that we referred to earlier.

SOURCE

Few sources except advertisers ever admit that they are engaged in propaganda. They usually call it information or education. To a President of the United States, only unfriendly nations practice propaganda, never a President who is doing his best on network television to persuade his audience that he is doing a good job, that we will soon be out of Viet Nam, that we have been wholly successful there. Any President is a propagandist when he says anything in any year of any administration. So are all his top officials. So are Governors, Mayors, and all other office holders. They may look as if they are simply answering questions from reporters—that is what they want you to think they are doing. They may look as if they are making an impromptu speech. In fact, they are carefully choosing words, slogans, ideas calculated to persuade the audience, including the reporters, to accept whatever opinion or interpretation the official has in mind.

A source may be a witness before a congressional committee. If the chairman asked him to appear, he will say things that support the chairman's views. If the witness asked to appear, he wants a chance to try to publicize his views or at least to try to persuade members of the committee. Any time a group, no matter its size, appoints a director of public relations, or information, or public affairs, a source is created.

Propaganda also comes from many sources besides those called propagandists. When a pastor preaches a sermon designed to persuade his audience to a point of view, he is engaged in propaganda, having chosen the phrases, arguments, and scriptures that enforce his argument.

The air would be clearer and citizens would be better able to know their own minds and to get what they want if all Americans would stop thinking of propaganda as sinister and realize instead that it is ubiquitous. A tough-minded Frenchman summed up accurately, ". . . nowadays propaganda pervades all aspects of public life." 6

WHAT: CONTENT

The content of any piece of propaganda can be taken apart word by word to reveal what its source had in mind. Thus when gathering intelligence from secretive nations, the United States and other open nations keep a close ear to the radio and read the press. A secretive nation, under strict censorship, puts out propaganda for its own people. Such propaganda often tells a great deal about what concerns the leaders of a secretive nation; even strong denials may indicate that what is denied has in fact happened.

Content consists of selected facts, selected ideas, selected opinions. These are expressed in symbols and slogans.

A symbol is a word or thing that is taken for the idea or thing with which it is associated. "Frontier" in America means the country to the west. "Liberty" has a meaning that can seldom be defined exactly but means more or less the same to all Americans. Few laymen to political and economic theory have a correct knowledge of what "communism" is, and few beyond specialists in Soviet government and economics know how communism is practiced in the Soviet Union; yet American politicians for twenty years after 1945 used "communism" as a symbol of threat and as a reason to wage cold and hot wars. Hammer and sickle mean Soviet Union and communism. (In the 1920's a bushy-bearded man carrying a lighted bomb meant "communist" or "Bolshevik.") Swastika means Nazi Germany.

Sounds can be symbols. When background music emits a short and sinister buzz, apparently borrowed from the rattlesnake, danger is near. Castanets evoke Gypsy dancing girls. Theme songs can become so widely known that they come to stand for their users. Only the President can enter to "Hail to the Chief." Tex Ritter singing "High Noon" from the sound track came to represent the entire movie of that name. "I Walk the Line," played in the low notes of a guitar, can mean only Johnny Cash.

Words and phrases are symbols. They are chosen very carefully for the purpose of propaganda. What Governor Wallace would call some pointy-headed intellectual chose the Dow Chemical Company as the target of student radicals' propaganda against the war in the late 1960's. "Dow" is short and fits headlines or placards; it is impersonal, flat in sound. "Napalm" symbolizes fiery death and destruction. Only a very small part of the company's production was napalm. The combination of "Dow" and "napalm" was perfect for sound, conciseness, symbolism, and all other requirements of radical propaganda. As a result, the company was severely hurt, and its officials were forced to spend much time trying to get an audience for their counter-propa-

ganda. "Intellectual" itself is also a symbol, whether pointy-headed
or not. Some others are "liberal," "conservative," "home-maker,"
"businessman." In any of these categories shades of meaning defy
any generalized description of a whole group.

Symbols often are euphemistic, designed to divert the receiver
from a harsh reality. When the United States was getting deeply
into war in Viet Nam, we heard only about "escalation" of the
fighting, not about entering a rugged struggle to win a major
war. After we stopped bombing North Viet Nam, any further air
raid across the border was never called bombing but rather "protective retaliation" and then, after someone thought up a better
combination, "protective reaction." Retaliation suggested re-
venge. One could watch the Secretary of Defense facing reporters
and television cameras and in a weird distortion of sense insisting
that our policy had not changed. We have not resumed bombing
North Viet Nam. We are sending bombing missions in protective
reaction against positions in North Viet Nam that shoot at American reconnaissance planes with armed escorts flying over North
Viet Nam. Such are the power and prevalence of symbol-speak
that reporters accept the distortion and transmit it to all who are
interested.

Military jargon is crowded with euphemisms as symbols, especially when the United States is not clearly threatened and the
military and civil politicians must keep citizens persuaded that
they should accept large expenditures and heavy casualties to
fight a dubious war or to maintain large forces on guard overseas. We are not in a "war" in Southeast Asia; we are granting
military aid. "Casualties" is a standard term in all wars to mean
dead and wounded. We do not invade another country; we make
an "incursion." An invasion to cut enemy supply lines is "inter-
diction." Everything from patrolling the earth, the sky, and space
to bombing one suspected location of one Viet Cong soldier is
done for the sake of "national defense" and the protection of the
"security" of the United States.

The President uses "Vietnamization" as a *package* symbol to
mean his policy of "withdrawing" American ground combat
troops (but not air combat forces and ground supply troops). It
is a cover-up symbol, an evasion of the more candid, "I am trying

to get out of a war we can't win under our decision of limited
military action. My predecessors made mistakes by getting involved, but I approved at the time. I am in office when a great
many Americans no longer approve of the war. They may not
vote for me next time unless I can convince them that I have
really reduced the numbers of Americans in Southeast Asia. I
only hope the South Viet Nam army, even with our air support,
our advisers, our supplies and our support of the country's economy does not fold and quit before I can get enough Americans
back home."

The propaganda for peace has some euphemisms too. "The
brotherhood of man" represents a concept that can be accepted if
one does not ask the bothersome questions about nationalism.
Some "doves" as opposed to "hawks" approve of "wars of libera-
tion" although others regard all war as a "waste of lives and
money."

Slogans are short statements of opinion which become known
to so many people that they are used in the same way as symbols
for communication. State a slogan and the details can be spared;
the idea is established. "If we don't stop the communists in
Southeast Asia they will be in San Francisco next." "America has
no interest in Southeast Asia." "The United States must honor its
treaty obligations!" "America needs to regain the respect of other
nations."

Surprisingly few symbols of propaganda have become the cur-
rency of communication in the movement to save the environment. "Pollution," "environment," and "ecology" are about the
only ones. "E-Day," when students and others across the land
brought attention to the question, was poorly chosen because it
was limited to one-time use. "Whole Earth" caught on with the
considerable but still relatively few numbers of people who
bought the catalog published by the Portola Institute.[7] The
Whole Earth symbol is broader than environment, however. It
includes a way of living cheaply and naturally by escaping from
the urban-industrial world of chemical foods and expensive
clothes, a manner of living that its followers would sum up in another symbol, "life-style."

HOW: MEDIA

The media of communication can be classified in a number of ways, reaching at times to such an extreme as the discovery, and worse reporting, of a category called memorandummatic.[8] Categories are a sometime useful thing.

Most important, when thinking of the media, is to remember that their variety is infinite. A new one will appear (the picture phone, computer pictures, video cassettes) as soon as technology and the market are ready. Attempts to find new uses are incessant and ingenious. A wise propagandist is always looking for them. For example, the electronic media are most familiar as radio and television. They also include inspirational speeches on tape, played solemnly at meetings of the faithful in their local cells, whether Birchers or Maoists, Christians or communists. They include telephone hook-ups to be received either by individuals at their phones or by audiences in auditoriums. Any communication by sound or image carried by electricity by any device can be used for propaganda. So can print in any form and words spoken face to face with one or a thousand other people.

The printed media include the obvious newspapers and magazines plus newsletters, direct-mail, and thousands of specialized magazines and newspapers for particular audiences in business; religion; the professions; farming; sports; hunting and fishing; conservation, subdivided into soil, water, air, health, mammals, reptiles, birds, forests, pesticides, fertilizers; and on as far as Americans form organizations, which is their tendency. Posters are printed; they range from a notice on a bulletin board to billboard advertising.

Some other media are neither printed nor electronic: speech in all its uses, for example, from spreading rumor to preaching, demonstrating, sabotaging, hunger striking, rioting, and others.

A key element is to choose those media that will best reach the audience whose support or acquiescence is most needed. One example will illustrate. Plant managers and mayors are more strategic in the immediate use of the environment than are college students. If quick action is needed, the media should be chosen to reach plant managers and mayors, perhaps their trade journals, or telephone conferences, or personal appeals by speech or letter. If more lasting reform is wanted, the appeal can be distributed next in the media that reach college students, for they will become the plant managers and mayors of the future. To reach the residents of a neighborhood, the best medium may be door-to-door visits. To reach the residents of a city, the best media may be newspaper, radio, and television. To reach an audience of the few top executive officials who make decisions for the nation, the best media may be a nationwide campaign in all media or personal visits from equally prestigious people. The choice of media can be decided only case by case, subject by subject.

WHOM: AUDIENCE

Only the amateur expects everyone who is exposed to his propaganda to be paying attention to it. The professional knows that only some will listen. These will be the people who already have an interest in the subject or those who catch the mention of it and become interested. They can be attentive yet still be for or against the proposed change. The purpose of propaganda, then, is to reinforce the support of those who are already interested and in agreement and to capture the support of those who become interested and might be persuaded.

Only seldom can an audience be defined neatly and easily. The people who come to a meeting to hear about a subject, the subscribers to a publication of narrow scope, can be defined easily. Most of the time the "source" sends the "content" to the "audience" that it thinks will most probably include the people who will pay attention. Within a category, however, sub-groups will be interested in differing subjects. Within the politics of medicine, for example, internists differ from surgeons, psychiatrists from internists, orthopedists from psychiatrists, salaried physicians (in research or public health) from fee-earning doctors, small-town clinicians from big-city medical center physicians.

Always within any category the young will differ from the "establishment." Changing ideas, changing moods, come from the fact that the young fortunately often come up with better ideas after having watched their elders make mistakes. The United

States of the 1970's is a vastly different place from the nation of the 1950's. A generation of the young has come along to take over.

EFFECT

To measure results precisely in a large audience costs too much for most propagandists to take the trouble. It requires survey research. Instead, most politicians watch for changes in their own way. A colleague changes his position, a friend calls to approve, votes (or sales in business propaganda) rise or they do not. Persistence, sometimes over a span of years, is required according to such circumstances as the novelty of the subject, the size of the change proposed, the size of the audience to be reached. If attitudes toward the environment are to be changed so that life can continue, the propaganda and politics must be heavy and persistent.

THE PRACTICE OF PROPAGANDA

The rest of propaganda is practice. It is learned from experience. A beginning group of environmentalists or an individual should seek advice from professional politicians and propagandists before planning a campaign. Practice varies from one cause to another. Sources, content, media, audiences, and effects differ. Certain practices are so accepted, however, that they can be called the wisdom of the trade.

The best propaganda looks so much like news that not even the reporters mention that it has a certain purpose. You can analyze the events of any day and see that many of them were made to happen: a certain witness before a congressional committee, a demonstration, a presidential news conference, an action in war that focuses attention on the plight of prisoners of war and away from continuing casualties and the war itself, an earnest statement by a President or a Secretary of Defense that turns out later to have been deceptive because it was not completely candid.

Not too many years ago the Governor of Alabama blocked the door of the University of Alabama to prevent the admission of a black student. This act insured: (1) live television and radio coverage; (2) prominent space in every newspaper; (3) news maga-

zine coverage; (4) international news space; (5) editorial comment; (6) replays in news summaries on radio and television and
in reviews of the year. His propaganda and political strategy was
simple: As Governor of Alabama I am opposed to the federal authorities forcing integration in our schools and university. I do
not have the power to stop them, but I will make them forcibly
move me aside to enter the building. I will have done my best to
abide by the feelings of those who elected me, and I place full responsibility for this action on the federal government. The Governor became famous and began to run for President.

Take advantage of breaks in the news. If coho salmon are
seized and destroyed by the State of Michigan because they are
too contaminated with D.D.T. to be fit for human consumption
and you are fighting the use of D.D.T., exploit the incident for
its news value.

Make news at the right time. If you plan a demonstration,
stage it at the right time of day for press and television coverage.
Know deadlines for afternoon and morning papers, for radio and
television news programs.

Know what is most likely to go on radio and television networks, and on newspaper wire services, and never, never ask a
news handler to use something which is not news. Drive home
the point by turning it into an event that is news!

Try for double effect. Governor Lester Maddox of Georgia is a
master of double effect. He demonstrates as a one man picket in
front of the Atlanta newspapers building, and the press and television networks give him full coverage. He claims to be angry
over something said on a late night television talk show and
walks off. Because no other guest has tried this stunt, he makes
news; he get his point across and continues in the limelight.

Do not be deceived by ardor and your own propaganda. A surprising number of politicians and others who use propaganda
fail to keep clearly in mind what is real support and how much
they are reading into something that is not really there. Especially when employing face-to-face propaganda in crowds, as in
parades or demonstrations, pay little attention to what is happening at the time. Watch television and press to see how the much
greater audience that was not present received the performance.

Student radicals in the late 1960's and the staff of the late Robert F. Kennedy in the primary campaigns for President in 1968 both made this mistake. The radical student leaders claimed they had support because they drew many students out to demonstrate. More students, and practically all off-campus citizens, were turned off by the stammering and unreasonable people that many young demonstrators appeared to be on television screens. The Kennedy parades drew:

> . . . ecstatic, screaming, jumping crowds of students, ghetto blacks, Mexican-Americans, and poverty-stricken Indians. The press and television covered the frenetic crowds on hand—and rhapsodized. At first, the Kennedy campaign staff was pleased, but they soon discovered something else: Millions of Americans watching television were appalled at what seemed to be the radical and frenetic nature of Senator Kennedy's campaign. His ratings in the polls fell, and soon the nature of Kennedy's campaign changed.[9]

Be aware of the mood of the time. Is it one of well-being or of concern over war, crime, income, the next generation? If people are troubled, what troubles them most? It appears that President Nixon and his Vice President misinterpreted the mood of 1970. They stressed the symbols "violence," "law-and-order," or the "social issue." The supporting *Wall Street Journal* in an editorial the day before election (November 2, 1970) referred to the social issue as trying to stop "a litany of what for want of better words must be called anti-Americanism and permissiveness." In the news columns of the same paper and in conversations all through the land, among all kinds of Americans, the first concerns were high prices, falling real incomes, sluggish business, lay-offs, and the fact that despite all the Nixon symbols about "game plan" and similar magic, times were not getting any better. The Republican Party failed to gain significantly in the Congressional elections of 1970 as it had hoped to do. Its leaders had failed to keep in touch with the mood of the time.

The most accessible readings of a mood, as well as of opinion on a particular issue, are the public opinion polls. Learn to read them. Analyze the questions carefully. They may not get the answer they seem to be getting. Make sure you know how much time has elapsed since the poll was taken. It is just as important to know what a poll does not tell as what it does. Valid conclusions can be based only on what the polls do tell.

Never underestimate or overestimate an audience. Not all Americans are the same. Only very few of the influentials fall for fads in ideas, books, hair spray, deodorants, and fashions. Perhaps there are some women like those who insult each other's wash, and some men and women like those who hawk pills, lotions, and gargles in television commercials, but they will not be among the people who successfully affect vital decisions.

Data about age distribution, race, income, things possessed, and occupations can be found in the nearest library from census reports. Ask for the reports by nation, state, county, or census tract, the last being the smallest geographical areas for which the data are reported. For a campaign aimed at individuals the data from a census tract is probably the best.

One does well always to remember that Americans of all ages and incomes are among the best informed and most widely experienced people in the world. They are exposed to a great array of information and propaganda. If they tried to absorb all of it, they would end up as gibbering lunatics. They listen to what interests them. Each group that wants to reach them has to find those in the great public who are interested in what it has to say.

The exposure of Americans to information and other experiences is progressive. Each generation that grows to maturity is better educated, more aware, better informed than the generation that begot it. Appeals for attention have to change with the changing abilities of each new generation.

Honesty is the best policy. It is the right policy. It is also the only safe policy. In the open society of free America, propaganda is competitive, and someone is always watching for a flaw in the other's propaganda. Professional newsmen, especially, want to handle only honest news. Their ethic requires that they make their stories as factual as possible. If they discover that a source has misled them, or given them only partial information, they will expose him. President Nixon was not helped after the elections of 1970 when he tried to interpret the results as a victory for his Party. It does not help a business firm to advertise that it has removed pollutants from the water beside its plant when reporters with cameras can show that it has not. Only the ignorant trust the Atomic Energy Commission to tell the truth about fall-out or any other nuclear danger; the agency has lost its reputa-tion with the informed and the skeptical.

THE AMERICAN PROCESS

Politics in America reflects two characteristics so prevalent that any citizen who wants to help save the environment will have to defer to them if he or she is to be successful. The American pro-cess is a matter of groups, and it is a matter of gradual change.

The capsule word that comes closest to describing the way gov-ernment, business, and other institutions in America make deci-sions is "groupism," which has a dreadful sound because the concept has been so abused by organizers and joiners. David B. Truman, another political scientist from Chicago of the 1930's, best stated recent group theory in *The Governmental Process, Political Interests and Public Opinion*, an enduring book.[10] Groups, according to Truman's theory, can be organized and carry a name or they can consist of people who share an attitude. A group of individuals who share an attitude makes certain claims on other groups in society, either by clear statement or by inherent, potential statement. When some development arouses all the individuals of a particular group, it ceases to be latent and becomes an interest group.

Truman's theory of groups struck most political scientists as a true statement of what happens in the process of government in the United States. Some groups are organized—a church congre-gation, for example. Other groups exist because their members share an interest, but these groups are not organized—"church people," for example. An unorganized group can still be a force simply because it exists, and "church people" are respected by leaders. When Calvin Coolidge was asked to explain the cause of the creeping liberalism of his time, he answered, "women." Women are another unorganized group.

In the day-to-day practice of American politics, however, the voices most heard will be those of organized groups with officers responsible for knowing what is going on. The potential interest groups have influence only because those who make decisions know they are out there and must not be offended—too much.

If a group does not organize and post a guard, decisions will be made without the knowledge of the members of the potential group. They wake to the news that they may be dying of pollution; a slice is torn from a park for a wider freeway; the estimated time of arrival of doom has been advanced again. The lesson for those who want to save the environment is plain. They should organize and hire a guard, a staff to keep watch against offenders, an information center to keep members informed, and a quarterback to call plays for the offense in legislatures, stockholders' meetings, local governments, state governments, national government, executive, legislative, and judicial branches, factories, retail stores, banks, insurance companies. Decisions about the use of the environment are made in all these places and more, and decisions can be influenced best while they are being made.

As for gradualism, the United States has been having a revolution for some two hundred years without once having the government overthrown or the economy drastically changed. Pragmatic Americans accepted the principle that government, business, all other institutions and individuals could change and that as long as the change made sense, it would be accepted. They interpreted the Constitution and the mixed system of private and public property as allowing for change.

Individuals and interest groups are perpetually seeking one change or another, depending on their interests and their values. When they command enough attention to start a movement, the issue is debated, discussed, amended, approved, or disapproved. So far, those who advocate drastic change, such as replacing private with communal ownership, have never been able to win more than a passing curiosity from most Americans.

The new movement to save the environment is being discussed and conducted at the very center of the American process. Its first advocates picked up many others who agreed with them, supporters of many varied interests, backgrounds, occupations. Legislatures have been persuaded in many places and on many subjects. Private and public bureaucracies have been moved. Yet nothing has been done, or will get done, abruptly or drastically.

The progress of change will be gradual. Today in one state, in one county, a decision will be made about local water. Tomorrow the nation will move to begin saving the oceans. Day after tomorrow in some place a decision will be made for clean air.

If the reformer becomes impatient because all problems are not solved at once and completely, he frets ineffectively and drops out of the process by which change is accomplished. He should, instead, take comfort in the fact that all the incidental changes do add up and that the process of constant change holds promise that change will continue. As technology changes, the nation will be able to protect the environment more effectively each year, and this too is an advantage of constant change. Indeed, in view of the American experience, the worst blow that could be dealt us would be a fast and total overthrow in favor of a frozen system that in the faith of its advocates is seen also as a perfect system.

NOTES

1. A recent example is Richard M. Scammon and Ben J. Wattenberg, *The Real Majority* (Coward-McCann, New York, 1970). Studies of voters using modern techniques, go back to Harold Gosnell, *Negro Politicians* (University of Chicago Press, Chicago, 1935), a Chicago study. For national elections Louis H. Bean had discovered in the 1930's that certain counties would reflect the national vote. See his *Ballot Behavior* (American Council on Public Affairs, Washington, D.C., 1940) and *How to Predict Elections* (Alfred A. Knopf, New York, 1948). Another pioneer study was Paul F. Lazarsfled, Bernard R. Berelson, and Hazel Gaudet, *The People's Choice, How the Voter Makes Up His Mind in a Presidential Campaign* (Columbia University Press, New York, 1948), a study of the 1940 presidential campaign and election in Erie County, Ohio.

2. V. O. Key, Jr., *Public Opinion and American Democracy* (Alfred A. Knopf, New York, 1961), pp. 8, 27–37.

3. Key, *ibid.*, p. 6. Leo Rosten, one of Lasswell's students, wrote an affectionate essay about him in *People I Have Loved, Known or Admired* (McGraw-Hill, New York, 1970), "The Incredible Professor," p. 274. The titles of Lasswell's books written in his days in political

science at Chicago were: *Propaganda Technique in the World War* (1927), *Psychopathology and Politics* (1930), *World Politics and Personal Insecurity* (1935), and *Politics: Who Gets What, When, How* (1936). Then he went east and became more polite. When three scholars published in 1971 their analysis of 62 basic innovations in social science from 1900 through 1965, Lasswell was the only man named three times as associated in the contribution of three different innovations: elite studies, quantitative political science and basic theory, and content analysis. Of the four men listed twice for contributions to basic innovations, two had been Lasswell's students or associates, Herbert Simon (hierarchical computerized decision models and computer simulation of social and political systems) and Ithiel de Sola Pool (content analysis and computer simulation of social and political systems). Karl W. Deutsch, John Platt, Dieter Senghaas, "Conditions Favoring Major Advances in Social Science," *Science* Vol. 171, Feb. 5, 1971, p. 450, Table 1.

4. Murray Edelman, *The Symbolic Uses of Politics* (University of Illinois Press, Urbana, 1964), p. 20, and all of this book.

5. Harold D. Lasswell, Daniel Lerner, and Ithiel de Sola Pool, *The Comparative Study of Symbols, An Introduction* (Stanford University Press, Stanford, Calif., 1952), p. 12.

6. Jacques Ellul, *Propaganda: The Formation of Men's Attitudes* (Alfred A. Knopf, New York, 1965), p. 119.

7. *The Whole Earth Catalog, Access to Tools* was published twice a year by the Portola Institute, Menlo Park, California, between Fall 1968 and Spring 1971. It described a great variety of things that were worth the price and that would help an individual be more self-sufficient. The Catalog was a great success.

8. James N. Rosenau, *Public Opinion and Foreign Policy* (Random House, New York, 1961), pp. 91–93. Memorandummatic media; e.g., a news letter, should not be confused with assemblematic media; e.g., a meeting, pp. 86–91, or programmatic media; e.g., an agenda, pp. 93–96.

9. Scammon and Wattenberg, *op. cit.*, p. 19.

10. (Alfred A. Knopf, New York, 1951). Truman drew from the ideas and observations of A.F. Bentley who could hardly have been less a group man in person. He lived in rural Indiana, thinking for himself, not conforming to the fad-think of any groups.

---
chapter
—— 9 ——

GOVERNMENT AND BUSINESS

»»»»»»«««««

GOVERNMENT USUALLY WORKS with business in the United States to handle nearly all the big jobs except welfare, education, and the postal service. Some citizens may think that the federal government alone fights wars and sends men to the moon. In fact the government fights wars with a technology developed mainly by business and purchased from business. For a trip to the moon, the government sets the purpose and provides part of the personnel. The rest of the venture is a joint affair with business.

Probably the greatest man-made structure in the history of the world is the interstate highway system, although how does one really measure it against the Great Wall or the Pyramids of Teotihuacan, considering the resources and technology available when these were built. Three kinds of governments—federal, state, and county or city—are involved in maintaining the interstate roads, but the construction is done by business with materials that come from business.

Government will work with business to save the environment. These two institutions, in combination, may be the most competent social instrument ever developed by man. Public purpose is defined by government. The profit motive is the incentive for business to join with government to achieve that purpose.

THE NATURE OF BUSINESS

The American system of business–government partnership is one of the least understood of social instruments. It is contrary to

the theoretical separation of public and private, profit and non-
profit, regulated and regulator, tax collector and tax payer. One
reason for the lack of understanding is that no general reference
to business will hold water. All the usual symbols, such as "a
business man" or "business," exist in the mind, not in fact. A
businessman can be the owner-operator of a drive-in eating stand
who turns every hamburger himself. Or a man in the same line
can own some shops and franchise many more in a nationwide
chain. Add some bedrooms and a larger menu, and he is in the
motel business, but still running a kind of drive-in. A dealer—
repair shop—service station in a blister town on the windy
plains of Texas is part of the automobile industry, even though
it is small and rundown. General Motors Corporation, the nation's largest industrial firm, is also part of the automobile industry.

Citizens seldom understand that business can be discussed
only one detail at a time. Some businessmen, but not all, will do
anything for a dollar (and incur the distaste of many other businessmen). Not all industrialists, acting as one, deliberately poisoned the air and water. They did the obvious at the time. Just
as city councils and farmers did, they used the land, air, and
water as the cheapest dump available. All industrial societies ignored the accumulating threat until imminent danger made
them take notice.

Not even the people who think of themselves as businessmen
understand American business. They are too widely separated by
rank and function to know one another. One group consists of
the lesser executives of great corporations. They head plants,
manage divisions, supervise construction or buying, or any of the
many jobs that must be done. They know the top executives only
by name and from hearing them speak at regional meetings. Also
in this group are the owners and managers of small companies,
perhaps a city-wide chain of cleaning and pressing shops or a single walk-in store. Associated with these men will be the salesmen,
insurance men, lawyers, and accountants who live from the services they render.

Such businessmen seem to spend inordinate amounts of time
reassuring each other that they are a special backbone of society,

truly endowed with virtue, and that such facts as government and taxes are a plot against them. Their talk becomes a ritual, almost like shaking hands.

Such talk would be embarrassing to most of the chief executives of the great corporations. They are much too sophisticated, busy, and alert to complain about what they, after all, can influence. They accept the tax structure because they have studied the alternatives. If they want it changed, they can suggest modifications to their friends in Washington and perhaps testify before a committee of the Senate or House. Many of the top few in business, or their lawyers, bankers, or brokers, have held similar posts in government. They know that no sinister plot exists. The state of business depends on the state of the nation, these men know, and the interests of business and society are the same.

Fortune, a magazine designed to appeal to top executives, keeps the record on them and their corporations in an annual summary of facts about the 500 largest industrial corporations and the fifty largest life insurance companies, commercial banks, retailing companies, transportation companies, and utilities.

The top executives of the top corporations are well educated, and they would not have reached their present posts if they had not been capable of thinking from facts, were honest, honorable, confident, and able to do well a variety of things. Forty-four per cent in 1970 held postgraduate degrees, another 36 per cent undergraduate degrees, 14 per cent said they had some college, and only 4 per cent replied that they had not gone to college at all. More of them attended Ivy League and other private colleges rather than state schools, and, to emphasize their uniformity, 80 per cent were Protestant, 80 per cent voted Republican, and 80 per cent had annual incomes in the range between $101,000 and $399,000. Their main interests outside work were a second home, travel abroad, art, and boating.[1]

These men are far removed from their lesser brothers in the organizations. They have their own clubs. They feel easiest in the company of one another. Some of them reach a state of personal and social assurance in which they look upon their automobiles not as evidence of their status but actually as something they like

to drive, whereas their subordinates in the corporation are careful
to match the makes of their cars to their positions. Some of them
pursue scholarly hobbies and write books. The world's best book
on hummingbirds was written by the head man of du Pont
Chemical.[2] He is also on the boards of three institutions of
music.

Another source of misunderstanding is the tendency in America either to be for business or against it. How this situation came
about is puzzling. The European class distinction of "those in
trade" could hardly have become strong in a new society without
an avowed aristocracy. Perhaps the distinction in America between farmers and workingmen on one side and businessmen on
the other explains it. The quarrel over economic benefits was
inevitable. Farmers resented freight rates, interest rates, and
insurance rates, and became the backbone of the populist movement. Workingmen resented low wages, long hours, and poor
standards of work, and formed the labor movement.

The nineteenth century business barons helped greatly to
prove their enemies right. They did exploit their victims, including one another. They did say "the public be damned." They
did charge all the traffic would bear. Some of them did commit
crimes against the law and against the opinions of mankind.
Some of them did join crooked politicians in greedy schemes;
and it was no excuse to say that it always takes a crook in government to serve a crook in business; they both get unmerited
gain, the business crook usually more than the political crook.

Intellectual leaders, the novelists and poets, some editors and
reporters, some historians, lost respect for business. An odd comment on American literature is that for such a central and vital
part of American life as business, only a few minor novels have
dealt with business from the inside. Our major novels deal with a
possessed captain of a whaling ship and a runaway who today,
sadly, would be put in charge of a juvenile court.

One effect of these developments was that Americans of "liberal" persuasion grew to distrust businessmen and to resent their
influence in social affairs, forgetting that all through history the
holders of economic power were also the holders of political

power. Even in this separate continent that became the Americas, the founders of the first great civilization, the Olmecs, got political power because they possessed good water and rich land.[3]

When government and business collaborate to save the environment they will be using the best social instrument available for the purpose. This will not please those who distrust business, but in any event there is not much that can be said about the American system that would please them. Americans built a political economy in which government turned to so-called private business for practically everything that involved construction, manufacturing, and general technological development. Public administration by contract with business firms is as well established as public administration by civil servants. Government turned to business aware that the money paid would end up as profits and wages in private pockets, and that only some of it would return in taxes to be spent for public use.

From the beginning government also served business in the faith that profits and wages in private hands were good for all. The automobile industry would not be what it is today if government had not built and rebuilt highways to suit the cars produced, first by changing to hard surfacing for inter-city and other rural roads, then by progressive changes as speeds and weights of cars and trucks increased. Airlines could not exist without public airports. Many, many fortunes would not exist if government had not decided to open the public lands to private ownership. Later on government grew aware that the so-called regulatory agencies were not the adversaries of the businesses they regulated but more like benevolent friends. Government in America has always worked with business and been concerned for its welfare.

THE DIFFERENCE

There is still a difference between the duty of government and the duty of business. Government must serve the general welfare as it is defined in one situation after another. Business must make profits and continue to make profits.

A few but increasing number of business firms now acknowledge that they too have a duty to serve the general welfare. The president of Arthur D. Little, a successful research and consulting firm, thinks that the social forces now running in America will bring social as well as fiscal audits of business firms as corporate executives become convinced that they must be accountable to their employees, their customers, and the community at large. Social responsibility will become as crucial to the survival of a business firm as profit. For each proposed new product, for example, he thinks the tests applied will be not only technical and economic but will involve such questions as: Is it really needed? Will it pollute or pile up when discarded? Is it safe? What are its long-range social consequences?

More and more of the progressive companies among the great corporations have appointed committees of directors or of management and employees to question the social meaning of company practices. The biggest giant of all, General Motors, went outside the automobile industry to appoint a specialist in air pollution, combustion, and thermodynamics as its new Vice President for Environmental Activities. It followed up by creating a committee of six distinguished scientists to advise it on all aspects of research "including in particular the effects of General Motors' operations and products upon the environment." [4]

No one was talking of displacing profits, only of showing more social responsibility in the making of profits. The trend meant an even closer identification of business with government, for if business recognized that it too shared a duty to the general welfare, a new element had been added to American political economy, and the duties of government and business had become more alike.

The interplay of these two agents, business and government, showing some tendency to accept the same goals but not entirely, determines the success or failure of the initial effort to save the environment in the United States. The keys are initiative and control.

INITIATIVE

In the American system the initiative to save the environment comes from government, usually after political action by groups of citizens. Business for all its ingenuity in technical and managerial change cannot or does not initiate the great, significant new enterprises. It is handicapped by too little money for the really large investment, by habit, and by the risk involved in pursuing a new idea when an old idea is still profitable.

The federal government initiated the computer age when the Census Bureau introduced the Hollareth high speed data sorting machine. Government ordered the first electronic computers. The federal government also began the atomic age when it put up the $2 billion required to produce the first atomic bomb. Government made possible the aero-space age in its advanced form, and when government reduced the manned-space program, the aero-space industry suffered the shock of being unable to convert easily to other forms of production. With the possible exception of the automobile, business alone has not started any enterprise that produced an "age." We never see a reference to the age of the copying machine or an age of detergents, both major recent new changes started by business.

For anything much to happen in response to initiative, government and business will both have to act. In the usual run of things government guarantees investments and orders to purchase for a reasonable period of time. Some variation on the theme of guaranteed profits over a reasonable period of time can be found for management of the environment.

Initiative is always difficult to isolate in retrospect, for its true origin is in an individual who becomes concerned and who suggests action at a time when the subject is beginning its time.[5] If the time is right and the sponsors know the techniques of politics and propaganda, the idea will spread.

The individuals who initiate ideas and who know how to succeed in politics and propaganda are the bright hope at this stage of saving the earth's environment. True, they are not yet thinking in terms of the whole earth. True, they are not yet thinking of total recycling, total monitoring, total control of population.

Like all the rest of us, those who initiate ideas for saving the environment think in terms of the symbols of their psychological histories, and these symbols do not yet include the concept of Buckminister Fuller's space-ship earth. Only a few of the scientists and engineers among the concerned think in big terms; most of them stop with stereotypes. Mr. Fuller once teased the faculty club at M. I. T. when he said in a speech that the scientists and engineers in that throne room of technology were still thinking in reflexes. They spoke, he said, of "the sun setting" and the wind "blowing." They detected nothing wrong when a man in Houston asked an astronaut over China, and below the feet of people in Houston, how were things "up there." [6]

But this is the way we think and act in the early stages of a new idea. Elaboration and sophistication come gradually. Later the new stereotypes encompass more of reality. Change more drastic than anything suggested today, or any day, becomes acceptable tomorrow. The all important and essential thing is that the idea of saving the environment has been accepted in its beginning.

INITIATIVE TO SAVE ENVIRONMENT

For those who have observed trends in America since 1945, the rapid acceptance of the idea of environment has been impressive. It almost seems sudden. One year, the advocates of conservation, organized groups and their journals, talked more to one another than to the public at large. A year or two later their subject was news. National networks ran special programs. Entertainers and weekly series took up the subject. City councils listened to proposals for change that earlier had been considered politically unfeasible.

The scorn of practical men for what they had said was a sentimental regard for nature, and an affront to their righteous privilege to bulldoze, slash, and uglify in the cause of profit and low public budgets turned under pressure to concern. One year the engineers and administrators measured everything by cost in dollars against benefits to be returned. The next year their more enlightened members were ready to accept social costs and social benefits if someone could devise a way to measure such unprecedented ideas. Next they will be ready to accept the fact that such social benefits do not have to be measured precisely so long as the majority in society wants them. Education and public parks were accepted as valuable long before accountants began to try to find ways to define their benefits in terms as precise as dollars.

CONGRESS WAS FIRST

Congress was ahead of the rest of the country, ahead of President Eisenhower, who was stubborn in insisting that control of pollution was not a matter for which the federal government should spend money, while not admitting that the states either could not or would not spend enough money to get the job done. Congress was ahead of Presidents Kennedy and Johnson in the early days of their administrations. Congress was ahead of the polluting industries whose leaders wanted no regulation and no national standards.

Congress had first recognized the problem of air pollution as early as 1955 when it appropriated $5 million for research, a relatively small amount as federal money is allocated. The sticky question of states' rights was not raised then. Instead, Congress was assured by the bill's sponsor that he did not propose to put Congress in control of the air. He was from California. A member from New York said that smoke from New Jersey harmed his district in Staten Island, sometimes killing crops of commercial flowers and vegetables. A member from New Jersey replied that studies made by New Jersey showed that air pollution usually originated in New York. These two knew that interstate regulation was indicated but they were alone in speaking out in Congress.

The next year, in 1956, Congress adopted an act to provide federal funds to help cities build sewage plants. It did not appropriate enough money, and it required the cities to put up too much in matching funds, but it was the beginning of the government's acceptance of the idea that it had a responsibility for the quality of the environment to the point of intervening in local affairs. The formidable barrier of states' rights and arguments that the states and cities ought to pay the cost of cleanliness began to disappear.

Within the next fifteen years Congress adopted measures to do all of the following:

Discourage billboards along the interstate highways,
Broaden the enforcement authority of the federal government over water pollution,

Begin federal enforcement of laws dealing with interstate air pollution,
Purchase more land for recreation,
Establish a wilderness preservation system,
Encourage the states to adopt standards of water quality that would be
     approved by a federal water pollution control agency,
Expand the amount of money the federal government would match
     with states and cities to cut pollution of streams and lakes,
Set federal standards for emission from automobile exhausts,
Authorize federal grants to cities for the construction of facilities to
     dispose of solid waste,
Initiate federal, along with state, regulation of signs along all high-
     ways that receive federal aid and regulate the location of junkyards,
Declare a strong intention to improve the environment and provide
     for executive responsibility to protect it.

The forces at work in Congress for and against all this legislation were the expected ones. Mayors and other city officials
wanted federal money. Conservationists wanted any changes that
would stop to any degree the destruction of the environment.
The representatives of state governments were against federal intervention in areas that formerly had been considered under state
control. Businesses affected by each bill also opposed federal authority. If they were to be regulated at all, they wanted the states
to do it because they knew the states would be more lenient.

As usual, the deciding forces were the individual members of
Congress, especially the chairmen of the strategic committees.
Their knowledge, their perceptions of the subject, their personal
traits and characters went into their votes. Congress is a band of
individuals, and the last thing one should assume is that each
member reacts only to pressure and his own selfish interest.

Another law was discovered. It was adopted in 1899. Some
thought that this old measure might offer stronger protection for
water than any of the later laws. The Refuse Act of 1899 requires
industries to get permits before putting any waste into water.
Fines, even imprisonment, are the penalties for violation, but
court injunctions are probably more effective. The Corps of Engineers issues the permits with the approval of the Environmental Protection Agency, a new agency created by President Nixon
with the consent of Congress.

Congress brought this remarkable sequence of acts to a summary in the National Environmental Quality Act of 1969. It de-

clares that the national government seek by "all practicable means" an environment that supports diversity and individual choice, recycles depletable resources, and preserves a "balance between population and resource use which will permit high standards of living and a wide sharing of life's amenities."

The act directs all federal agencies to integrate the natural and social sciences and the arts of environmental design. The agencies must also take concern for the unquantified amenities as well as for economic and technical measurements. They must include in every recommendation for action affecting the environment a detailed statement of the expected effects, and they must make this statement public along with the comments of other agencies.[7]

None of these laws satisfied wholly its advocates or its opponents. They were adopted in a sudden rush of conscience and consciousness, but the turn had been made. In 1960 President Eisenhower had vetoed a bill passed by a Democratic Congress to expand federal aid to cities for new sewage plants. Ten years later a Republican President and the Democratic leaders in a Democratic Congress were competing for public credit for doing something about saving the environment.

Richard M. Nixon, who was a fundamentalist Eisenhower Republican when he was Vice President in 1960, became an environmentalist as President. In February 1971 he sent to Congress the most comprehensive message yet issued, in which he challenged that most sacred of sacred cows, land use. *Newsweek* could call him "the most engaged conservationist to occupy the White House since Theodore Roosevelt." [8]

He proposed regulation to prevent or control refuse dumped in the oceans, strip mines, the location of power plants on badly chosen sites, harmful pesticides, noise, uncontrolled development of open land, and the destruction of architectural treasures, as well as the pollution of air and water. His proposed steps were firm. For one, a federal charge would be collected from plants whose smokestacks emit sulfur oxides. For another, cities would collect from industrial polluters the cost of treating their wastes. A limit would be set on the noise from machinery used in transportation, construction, and other industries. The message warned the ecological banditti of continued strip mining: federal guidelines would be set for state regulation and if the states failed to set or enforce sufficient standards, the federal authority would enter the picture. For land use, the President recommended a National Land Use Policy to control development and to protect the environment. Some revision of economic incentives in the use of land would also be in order.

The downhill race of deterioration may have been stopped by acts of Congress and the promises of the President. We may have started the long pull to save life on the earth and to take our place in the Third Revolution.

CONTROL

Control has two meanings in the management of social affairs. One meaning covers all the techniques of knowing what is happening and what can be made to happen with the resources at hand. Accounting is the oldest form of this kind of control. In addition, this kind of control includes charts that tell a manager the location of all the parts that must come together in production, all the steps that have been taken, the place he occupies in a time schedule, and anything else that the ingenuity of man can think might be needed.

Some think that because a computer can handle so much information, its users tend to ask for too much, including some that is not really needed and that only adds to the paper work to be handled. Properly used, the computer has given mankind the ability to plan programs and to analyze operations better than in any other age. Applied to the use of the environment, proper control and use of complex information is a blessing never possible before to this extent.

The other meaning of control is the use of authority, and in nearly all cases concerning the environment this means the authority of the state. Law, power in the sense of authority to stop certain practices by imposing punishments and taxes and bestowing incentives, are the usual techniques of control and they are all within the authority of the state.

SOVEREIGNTY AND TOLERATION

Because control by law is the most likely way to control both private and public users of the environment, a reminder of the nature of the power of the state is useful here. Many people seem to assume that "government" could stop all pollution and deterioration if it would only act. Actually, government, or more precisely certain individuals in government, is only one participant in politics and propaganda. Other individuals and groups inside and outside government can also use politics and propaganda. Sovereignty to make law is not the whole story.

No government, unless it wants to spend most of the state's resources on police enforcement, may go beyond what its people will consent to obey. Nicolo Machiavelli, who was hardly a participatory democrat in *The Prince,* wrote several chapters of that book on how a ruler should conduct himself to hold the respect and favor of his subjects.

When we start the big push to save the environment, control through sovereignty will have to be of the kind that is tolerable to most citizens. The agents to carry out the measures adopted will be government and business, as usual, advised and prodded, as usual, by special interest groups.

The first great spur to initiative will be profits. In some cases a proper use of the environment will make profits; recycling of certain materials has already proved to be profitable. The second spur will be incentives, perhaps subsidies, perhaps reductions or refunds of taxes. The first is a matter for business, the second a matter for government. Both devices have been used before.

FEDERALISM

One of the most visible consequences of the Third Revolution as it takes place in America is a reorganization of the relations among the units of government. In the beginning separation was the rule in American federalism. Each state was jealous of its independence from the national, or federal, government and zealous in exerting its control over local governments, the counties and townships which then were the principal units of local government.

By the mid-twentieth century federalism had broken down. The states were not admitting what had happened and few citizens seemed to care one way or the other. Politicians at all levels of government still paid lip service to the ideal of federalism, but state and local governments were confronting a severe crisis. They were unable to control crime, spend enough for welfare, maintain schools at their recommended levels, or add and maintain streets, airports, and other services.

Within a few days the President recommended that federal money be given to states and cities to help meet their costs, the mayors of New York and Newark were quick to warn of the chaos that would visit their cities if the money were not forthcoming, and the governors of New York and Wisconsin appeared before a committee of Congress to tell of the states' distress, and to plead for funds.

Money, or lack of it, caused federalism's demise—not enough money to keep the states and cities performing their customary duties. Because of military and other costs added to the expense of conducting government through the years, the federal government loaded taxes on personal and corporate incomes so heavily that those states which already had adopted an income tax of their own feared to raise rates on already overburdened taxpayers and those states that had not yet levied an income tax were reluctant to add it. Instead, states resorted to sales taxes, and increased them. Cities and counties rely on taxes on real estate and personal property. These levies had been increased as much as local politicians thought possible. In some localities taxpayers refused to pay any more; they voted against bond issues for schools, and some schools closed.

Welfare was a disgrace. Crime was a disgrace. Poverty was a disgrace. Drugs were a disgrace. Slums were a disgrace. The only functions of government that were thriving at the beginning of the 1970's were war or defense and highways, the latter subsisting on income from their own taxes. Even spokesmen for these interests were crying for more. The American dream, the Great Society, all the other fine promises were turning into nightmares.

Historic arguments about states' rights had all but disappeared. The cities turned in desperation to the federal government for funds, while other local units that still could tolerate added taxes or higher rates tried to impose them. The answer, however, had to come from President and Congress if it came at all. Argument shifted from states' rights to questions of whether Congress itself could raise any more money to share with the states and whether the states should get it without Congress telling them how they could spend it.

The fiscal crisis, and the consequent crisis it evoked in the provision of governmental services closest to all citizens, meant another step in the reorganization of government, not by changing the Constitution but by changing practices within the framework set by the Constitution.

Local administrative decisions had proved to be a good idea in some cases but not in others. They were sound when they represented intelligent interpretations of state and national policy to fit a special circumstance. They were unsound when a township board could hasten eutrophication of a regional lake to the damage of many for the favor of a few. Citizens were no "closer" to state and local government than to national; not as close, according to Mr. Edelman. That outdated argument for a federal system was now irrelevant.

Some division of authority between federal and state and between state and local governments will continue. We have been evolving our present system of federalism since the beginning of the nation. Congress had first given land to the states to be used for education, canal and river improvement, wagon roads, and railroads. The first federal aid to higher education came in 1862 when each loyal state was given 30,000 acres for each senator and representative in Congress to establish "land-grant colleges."

Money began to follow land in 1887 when Congress extended subsidies for the establishment of agricultural experiment stations. In 1914 federal grants helped states create the agricultural extension system; in 1916 helped improve rural roads; and in 1917 helped establish pre-college training in agriculture and the trades. Federal aid in welfare began, not with the emergency of the Great Depression, but in 1921 when a federal grant was made to promote the health and welfare of mothers and children. The program lapsed in 1929 but was renewed under the Social Security Act in 1935.

The Depression and New Deal put the federal government into welfare in many respects, from jobs for the unemployed to slum clearance and better housing. Social security provides old-age and survivors' insurance; it also includes grants to states to help the old, the destitute blind, homeless, crippled, and dependent or delinquent children; and to assist programs of public health, vocational rehabilitation, and the care of mothers and children.

Most of the grants were to match state money for the specified purpose. Congress spelled out the terms in each act. The tone was that of a benevolent larger government that would help its poorer relatives as they helped themselves. Then the money ran short. The dollars spent were more than ever but they did not buy as much because costs were rising faster than income. Services declined.

N O T E S

1. *Fortune*, May 1970, pp. 180 ff. The next issue, June 1970, gives the same record for the second 500 industries.

2. Crawford, H. Greenewalt, *Hummingbirds* (Doubleday, for the American Museum of Natural History, New York, 1960).

3. Michael D. Coe, *America's First Civilization* (D. Van Nostrand, New York, 1968), pp. 123–27.

4. *The New York Times*, Feb. 14, 1971; *The Wall Street Journal*, Feb. 26, 1971.

5. Observers who write about how decisions of policy were made do not try to find the initiating individual because it is not that important to know. The statement that some individual starts it is based on common sense. Some excellent studies of policy decisions are Stephen K. Bailey, *Congress Makes A Law*, (Columbia University Press, New York, 1950); Raymond A. Bauer, Ithiel de Sola Pool, and Lewis Anthony Dexter, *American Business and Public Policy: The Politics of Foreign Trade* (Atherton Press, New York, 1963); James L.

Sundquist, *Politics and Policy, The Eisenhower, Kennedy, and Johnson Years* (The Brookings Institution, Washington, D.C., 1968); Townsend Hoopes, *The Limits of Intervention, An Inside Account of How the Johnson Policy of Escalation in Viet Nam was Reversed* (David McKay, New York, 1969); and a monumental study by Charles McKinley and Robert Frase, *Launching Social Security, A Capture and Record Account* (University of Wisconsin Press, Madison, 1970).

6. R. Buckminster Fuller, *Utopia or Oblivion: The Prospects for Humanity* (Bantam, New York, 1969), p. 12.

7. *Science*, Vol. 167, Jan. 2, 1970, p. 35. The record of Congress from 1953 through 1966 for bills concerning the environment is traced by Sundquist, *op. cit.*, Chapter VIII. The record after 1966 is taken from *The Wall Street Journal* and *Science*, various issues.

8. *Newsweek*, Feb. 22, 1971, p. 23.

chapter

—— 10 ——

GOVERNMENT AND BUSINESS: THE PROBABLE FUTURE

»»»»»»«««««

In a time of national crisis many propose, a few predict, and no one is certain. A few watch the pattern that develops from the perspective of past and present events. These few, however, can promise no more than conjecture. Change is unpredictable except to say that it is inevitable, and projection is no better than the assumptions from which it took off. Violence to preserve the past can delay change. Constant wars and military garrisons spread around the world consume too much public money and deny sufficient amounts to the task of improving life. Some new development, perhaps a low-cost, high speed, benign automobile engine or effective cleanser of smoke may make us forget that we still need total reform in our management of the environment if we are to save life on earth.

ASSUMPTIONS

Two assumptions precede the conjecture here. The first is that all the present steps to save the environment are only the beginning; more will follow. Too many people became aware and alarmed during the 1960's for this subject to be forgotten. They will continue to press for change, and their children will into the future. The second is that as the concern for saving the environment continues, citizens will tolerate more and more control over their lives. A catastrophe will make firmer these two assumptions

—sudden death for 100,000 people suffering under a five-day inversion over some large city, a traffic tie-up in which 5,000 die of carbon monoxide, a strange poison that kills 50,000 high school students because of something they ate in school cafeterias. We are not guaranteed exemption from any of these tragedies considering air pollution, traffic on holiday weekends, and the mass processing of food.

THE PROBABLE FUTURE: DESIGN

We are now set in trends that lead to a future that will not be very different from what we have now. Government and business will continue to work together. Nowhere and at no time has man found a better way of sharing goods and services. A combination of public authority and purpose with private profit and incentive gets done whatever the makers of decisions want done.

The national government will continue to work with state and local governments. To do this is not only the easiest course to follow politically; it is also efficient when done properly. The waste now is not caused by the way the work is distributed but rather by the failure to distribute the work because each government wants control of detail. When all three governments have to agree, or not agree, in numerous conferences about how to proceed in small matters, time is wasted. A clear division of jobs, with delegated authority to do each job, would be an improvement.

More and more the national government will set policy for what should be done about the environment. Not much happened in the present effort to save the environment until Congress acted. Any further actions will require federal money. Any further actions also will be fought—by some industries, some cities, some farmers, other interested groups. Congress is in a better position to handle interest groups than is a state or city. Most members of Congress come from mixed constituencies. A state legislator or city alderman is much more likely to represent a concentration of manufacturers, workers, blacks, whites, or ethnic Americans. The closer his constituency and the more uniform it is, the more a public official must defer to his constituents. A small city dare not tell a huge factory to stop polluting the air and water. Most of its citizens probably work in the factory. But federal officials can issue a stop order and then allow the city to enforce the rule.

The very nature of environment calls for centralized decision making. Very few things are local anymore. The condition of the air and water, of noise and strain, or ugliness seldom affect local interests only. A heavy smoke over St. Louis borne by southerly winds becomes a pall over Madison, Wisconsin. The outpouring of smoke along the Ohio River dumps particulates on Steubenville but spreads gases much farther around the globe. Air pollution over Los Angeles threatens to kill the oldest living trees on earth, the bristlecone pines in Death Valley 150 miles away. One settled point is that technology has brought us all closer together. Gary, Indiana's problems are our problems. Airplanes, automobiles, radio, and television have made us inescapable neighbors.

Only the federal government can make rules for the environment that cross local, state, and international lines. Only Congress has the power to regulate interstate commerce in the broad definition that the courts have given that term. Only the President, in some cases with the approval of the Senate, can make agreements with other nations. Only the federal government can afford to pay the cost of research and development to discover how serious the dangers are and how to defend against them. Only the federal government can pay the cost of cure. Only the federal government is remote enough from local and particular pressures to make the broad and severe rules that will require total recycling, total monitoring, and total control of population.

But the administration of the rules will be local. The geography of environment also requires this. While the consequences of violation are national and global, the offenses are committed one at a time on one farm after another, in one factory after another, at one city sewage plant after another, by individuals in localities.

Administration is now decentralized by two principal methods, each with variations in the amount of authority from standards set at the center. In one, the central government allows states and local governments to apply a central policy; in the other the central government itself applies the policy through its own regional, state, and local offices. Few Americans seem to realize that the federal government is all around them and not confined to Washington, D.C. They can find the evidence under the entry "United States Government" in the telephone directory for a city of any size.

Yet a third method of decentralization appeared in the administration of space technology. It reached its latest stage in the Apollo moon program and it may be truly significant insofar as the future of the environment is concerned. The government agency let contracts for design and production of the machinery and for the management of some facilities and programs. Then a government man and a private industry man worked side by side in each plant on each detail of the program. No longer was a government man barred from an industry conference nor an industry man from a government conference. Differences faded before the feeling of joint purpose and joint responsibility for the success and safety of the program. All the participants were kept informed, and coordinated, by a system of thorough communication in writing. Nothing was left to chance so far as human ability allowed.

When astronauts left the earth for their trips to the moon, they followed a detailed, step-by-step program of procedures, a copy of which was in the hands of everyone on earth who might be involved. When an oxygen tank accidentally exploded aboard Apollo 13 far out in space, the industry men were as concerned as the government men.

The Apollo experience may lead the way to a new form of administration which can handle the job of cleaning up the environment better than anything we have known before. It may lead to open recognition of an American transition from free enterprise to joint enterprise and from an emphasis on the policy power of government to an acceptance of cooperation between government and private agents. The new form of administration preserves the profit system but makes government the principal customer, and prices and profits become a matter of agreement, a condition already reached in the early 1940's in the production of military goods.

Some truths must be accepted before the new administration

will work for management of the environment, however. The purpose must be clear. We are just as determined to clean the environment, thoroughly, over a specified number of years, without yeaing and naying and changing our minds, as we were determined to reach the moon. We are willing to spend all the money necessary. We will pay the cost of the best talent to be found, the best equipment, the best procedures, and the best research. Even when the price is high, we will pay it, even though paying will mean a reduction in some other expenditures, such as those for war and defense and highways.

We accept the principle that in the United States the interests of government and the private economy are the same. We have already accepted this principle in practice but not in philosophy. Our political philosophy still assumes that government and business are separate. The Apollo practice assumes that government defines the purpose, makes the rules, and government and business cooperate in achieving the purpose. Government supplies the money; business does much of the work; both act side by side at all times.

We accept the essential role of government as definer and defender of the general welfare. Whenever any act threatens to harm the general welfare for the profit or comfort of one or a few, the government has a duty to prohibit that act.

We acknowledge that private motives, especially the profit motive, have not suddenly changed to one of pure benevolence and that many will continue to try to cheat, even when they acknowledge that the interest of business is also the public interest defined by government. Government will still use the police power to arrest, try, and punish offenders.

Surely the technology and management for saving the earth's environment are as simple as the technology and management involved in going to the moon. Surely these are most competent in the United States, the first and only nation so far to put men on the moon. Surely the United States can set an example for other nations of the earth. What we need next is a decision of purpose by the federal government and the acceptance of that decision by private business and all the rest of us.

THE DECEIVERS

The only healthy attitude at this stage of our journey is skepticism. Man has not suddenly turned rational. All men have not decided to be truthful. Plenty of us are deceivers. Some of us deceive ourselves. We advocate clean air but vote down bonds for improvements in mass transit that would take automobiles out of the central city. We assume that laws cannot be changed. We say that we will save the environment but we refuse to admit that we may have to give up some habits to which we are accustomed. We like to think that someone else is always the polluter, industry or a slaughterhouse, mining or airplanes, not recognizing that every last man, woman, and child is a polluter, sending all kinds of waste into sanitary sewers that are seldom treated enough and into storm sewers that are treated not at all. Charges to pay for thorough treatment plants, we think, are a waste of money; they don't buy things that can be seen and put in the picture window.

Others deceive all the rest of their unskeptical fellowmen. Some of these get caught when they speak with forked tongues in advertising. *Newsweek* once rounded up some double-talk deceivers with corporate big names. "It cost us a bundle but the Clearwater River still runs clear," said the advertising of Potlatch Forests, Inc., under a photograph of the river fifty miles above the company's pulp and paper plant at Lewiston, Idaho. At plant site both the air and water were incredibly polluted. The company had installed a car wash so that employees could try to save the paint on their automobiles from corrosive sodium sulphate in the air.

Southern California Edison, showing a picture of a healthy lobster, claimed, "He likes our nuclear plant." A marine biologist replied that the ad agency had borrowed the lobster from his tanks to photograph and that thermal pollution could be dangerous to marine life. Standard Oil of California praised its effort to reduce automobile smog and ran a picture of the "Chevron Research Center." The building was the Palm Springs County Courthouse. A number of critics, including the Federal Trade Commission, accused the same company of misleading advertising in its claims for smog reduction.[1]

Union Carbide, with one side of its corporate mouth, advertised its efforts to reduce automobile pollution. The other side for fifteen years had fought federal and state efforts to make the company clean up its plants in the Kanawha Valley of West Virginia. Some other corporate admen talk of particular efforts to reform as if they were universal practices, not mentioning that one or more plants have been cited by state and federal authorities for pollution. Plain doubletalk as transparent as these examples can be spotted easily by society's watchmen, and can be denounced as immoral if not also stupid.

More subtle is the fancier deception practiced by scientists who act as propagandists, and by other criers of untruths. Scientists are alarming when they become politicians. They seem to forget all the rules of evidence by which they practice their science and to propagate any kind of nonsense that they think will advance their cause. Some of them did this in the arguments over radioactive fallout in the 1950's. Some of them do it now in the arguments about environment. Pollution is a fact, and it is also a symbol with negative connotation. The scientists acting as propagandists who accept any kind of assumption or fact without testing are as culpable as the corporations which deceive in advertising.

The great truth is that all life on earth is dependent on the environment of earth. That environment cannot be used for an indefinite time as a dump. This truth argues that man should be very careful, but it does not give all the answers about how much care he should take.

A great deal of research needs to be done before a balanced use of the earth's environment can be planned in detail. We still do not know, for example, how many cubic tons of what kinds of waste can be oxidized naturally in what period of time. We do not know for sure what the effects of thermal pollution of water will be on what kinds of fish and plants. We do not know the final answers to how much dust the atmosphere can carry before we lose the sun. We do not know how many people the earth will safely and happily carry before population must be controlled. When propagandists talk only in absolutes about the environment, they mislead. When they are scientists, and talk this way, they are faithless to their training and ethic.

All good men know that the industrial nations have suddenly

discovered danger in their practices and that they should take drastic steps to reduce pollution while they carry on the research that will tell them how to use the environment wisely. Wisdom tells us to stop before it is too late. That is all we know, and all we need to know for now.

FINANCING

All the arguments, anxieties, and competition come to focus in the decisions about how to pay for saving the environment. And nowhere are the myths and stereotypes of American theory more rock-hardened. Charges are never seen as a distribution of cost but always as an attack on the purse.

Consumers, who are also citizens, never see prices and taxes in the same light. Their payments for streets and policemen are somehow different, in their minds, from the amount they pay for gas and electricity. When industry pays the bill for reducing pollution, consumers appear not to know that the increase will be passed on to them each time they buy paper, automobiles, gasoline, electricity, synthetic fibers, or pots and pans. Citizens when they pay taxes seldom realize that they are buying protection, education, highways and streets, armed forces, pensions, and all the rest.

The devices by which government can persuade, press, threaten, punish, and induce citizens to reduce pollution are limited only by imagination and the tolerance of the culture. They include rules and penalties, permits, inducements through tax refunds, subsidies, or advantages derived when government buys its own supplies. In 1971, as one example of the last, the federal government bought more than half its paper from plants that recycled paper. The devices also include charges for the use of a public resource, one of the best known being the effluent charge to be paid by a company that dumps waste into a stream.[2] They include fines and orders to cease and desist on pain of being closed. Again, a surprising number of small news stories appear in the business sections of newspapers about decisions against firms guilty of polluting. Such decisions are reached in the old slow way of hearings and appeals, and they may not be sufficient if peril becomes more imminent. But they are a beginning.

The easiest way to cut through the myth of stereotype is to continue to pay the cost by a combination of private and public finance. If anything is politically hopeless, it is a thorough reform of taxes in the United States. Taxes have accumulated through the centuries until they overlap, produce vested interests, provide windfalls occasionally, penalize some at the expense of others, and provide employment for untold numbers of accountants, lawyers, and economists, who spend their lives thinking about taxes. Public spending is easier to change. Legislatures decide how to spend public money all the time. More and more, as federal money goes to state and local governments, Congress becomes the one legislature that can control public programs by adding or subtracting from its grants. When the time is right to spend funds heavily for the environment, the principal decisions will be made by Congress.

With the exceptions of military and foreign affairs, atomic energy, postal services, space exploration, and a few minor functions such as the recording of patents, all of which are federal, public spending in the United States has become inter-governmental. Spending is decided and administered by federal, state, and local governments. The purposes for which governments now spend money, and for which most of us consent, are the best sign we have of what we think important. Spending by all governments in 1966–67 went for the following purposes by rank.[3]

Military, foreign affairs	34.4%
Education	18.5
Highways and streets	6.5
Interest on general debts	6.2
Natural resources	4.7
Welfare	4.4
Hospitals	3.2
Postal services	2.9
Space research, technology, and exploration	2.5
Health	1.2
Housing and urban renewal	1.1
Air transportation	0.6
Social insurance administration	0.6
Other	13.3
	100.0%

If these categories seem to omit some of the traditional functions of government, such as tax collection or law enforcement, there is a reason. These traditional functions have grown, of course, but not so much as the functions listed above. The category of "other" includes some important but no longer principal functions.

Similarly, some of the categories that are listed can be misleading. "Natural resources" means only in part attention to the environment. Of the 7.8 billion spent on "natural resources" by the federal government, 3.5 billion went for subsidies to farmers for "stabilization of farm prices and income." The $2.7 billion devoted to soil and water includes money spent for the development of electric power dams and plants, which often threaten the environment. Any item in a public budget has to be examined for meaning because the propagandists select the words, or symbols, to make the spending sound as acceptable as possible.

Despite ambiguity in the terms used, a budget is the most revealing document about government in America.[4]

The total in dollars spent by all governments for all the above purposes was $216.8 billion in 1967. As far as we can estimate, the amount spent by governments on the environment—then defined to include only air, water and, to a small degree, pesticide pollutions—was $1.2 billion. Industry at the same time was spending an estimated $0.9 billion,[5] yielding a combined estimate of $2.1 billion. The figure was not quite 1 percent (9/10 of 1 percent) of the total outlay of all governments for all purposes. It was barely a beginning.

At this point in any discussion of public budgeting, citizens, Presidents, and members of Congress usually ask where cuts can be made. The more explicit make suggestions. Cut the military. Cut out all waste. Cut the fat, whatever this means. Our politics seldom work this neatly. Every item within every item of every public budget in America has its special pleaders and defenders in private interest groups, legislatures, and executive agencies. Legislators hear from very particular interests back home. Officials in executive agencies know where their friends are and how to use their help. Some of their friends make every cartridge case and shoe-lace used by the armed forces. Others sell the seeds and silos to farmers who receive subsidies that hold up prices. Others are the leaders of unions whose members have jobs in the plants that make cartridge shells and shoe-laces and build silos. Should a reduction in funds be signaled by a change in policy, executive officials and their friends are quick with ideas, documented with statistics of need and promise, to spend the money that might have been saved.

Thus after the President's policy became one of reducing American troops in Viet Nam, the armed forces were quick to compete with one another and with civil agencies in ways to spend the "peace dividend," as someone called the prospective saving. The air force wanted a new bomber, this one to sneak in low and avoid detection by radar. The navy had a long shopping list of destroyers, carriers, submarines, and aircraft. The army argued for a new attack helicopter, a new surface-to-air missile system, and various automated devices for the battlefield. The President and Secretary of Defense renewed the talk of strong deterrence, both nuclear and conventional, as man's best promise of peace.

All the services were united in a strategy to reduce troop strength (no hardship unless there is war) but to put the money saved into hardware, for cannon, missiles, and all the other physical apparatus of war. *The Wall Street Journal* reported this intention to keep expenditures high and revealed the military propaganda line in the process:

> While troop reductions provide dollars, defense planners insist they also make the modernization effort all the more essential. "We must compensate for [the army's] smaller size with higher quality," says Gen. William Westmoreland, Army Chief of Staff. "We simply cannot risk both a small Army and an ill-equipped and undermanned one." Another general puts the prevailing Army sentiment more bluntly: "If we're going to be lean, we've got to be mean. What the hell's the sense in having an Army if it can't scare anyone?" [6]

The strategy could be attributed to any other set of executives in any agency in any government. They all have arguments for spending money saved in one program on another program within their jurisdiction.

Only occasionally is government spending reduced by much. Officials and the press often talk of cutting budgets. They mean that requests for money have been cut. A public budget is a request by the public executive branch to the legislative branch. Requests can be cut while expenditures can still be higher than the preceding year. The amounts appropriated by legislatures in response to budget requests have been rising over the long span. They have been deemed by legislatures to be necessary for new technology, new programs, new emphases in old programs.

Some programs, despite the trend, get less money. One current example is the space program, which has been reduced each year since 1966. The military program is reduced between wars. Over the decades, however, most programs, while they may fluctuate from year to year, will show over the long span that their expenditures increase.

New spending to save the environment will be for new technology and new emphases. We have had programs for a clean environment since water purification and sewage treatment began. Now we will spend more for better treatment plants, different automobile engines, cleaner airplanes, and all the new programs for making life more agreeable.

Funds for the new spending can come from all the various sources they already come from. If from private spending, they can be charged to consumers in the form of higher prices. If from public funds, if the mood is right and legislatures are worried enough about finances, the money may come from a reduction in appropriations for other programs, although experience indicates that this source is the least likely. Public funds for new spending can also come from deficit spending, the phrase that substitutes for saying that government spends more in a year than it takes in and so goes into debt. They can come from a constantly expanding economy which means that, holding to the same rates of taxation, the amount brought in by taxes will rise each year. Or they can come from new taxes or fees (a license fee perhaps to use the environment) or from an increase in the rates of taxation. Our problem in saving the environment is not money. The nation can pay the bill.

How much? Estimates occasionally appear in the news of what

it will take to clean up the cities, the water, the air. They are all guesses and they never answer the big question of how clean for how long. Instead of estimates of cost, all of us—citizens, legislators, judges, professionals, and private and public executives—will help themselves to understand costs if they accept some premises.

The first is that the clean-up cannot be done in one fiscal year. Financial planning should be for a very large expenditure over five or ten years in the first desperate effort to stop calamitous deterioration. The second period will come after the decline is slowed to a point of relative safety. It will be a time of steady and fairly large spending to finish the job of cleaning-up and to assemble the machinery and develop the practices to keep the environment stable. The third period will involve a steady cost for monitoring and enforcing all the new knowledge and new rules necessary to preserve life. For the United States there is a new emphasis, which hopefully will become one of the major costs of government.

At some stage Americans should make available what we have learned about management of the environment to all the people of the earth who do not yet know as much as we will have learned. If we are wise next time, we will not attach any money whatsoever and no military supplies and missions to the advice. The international program should cost very little if we forbear foolish gifts and corruptive military strings.

We will all feel better about the spending if we publicize the benefits that follow. Some of these can be stated in dollars—for example, the money to be saved when buildings and crops are not damaged by air pollution or the amounts to be saved when hospitals do not have to care for so many patients ill from foul air or water. Simply to declare such amenities will render them acceptable.

Certainly not much relative to the total, is being spent for saving the environment in America by any type of government. The biggest items are military and foreign affairs and education, the first taking 34.4 percent of total public spending and the second taking 18.5 percent. After them, but 12 percentage points below, come highways and interest payments. And 30 percentage points below military and foreign affairs, "natural resources" finds a place, but almost half of that amount really goes to farm subsidies.

So long as an advanced industrial nation such as the United States does not commit suicide in a nuclear war or die a slow death from pollution and continuous conventional wars, it can pay for cleaning up the environment. This is the important point. The nation's leaders need only to choose to give the environment a higher priority. Politics produced the distribution of public spending we have now. When the leaders agree to move, and the time is right, politics can rearrange the ranks of priorities. And the American system of business and government working together can carry out the programs.

NOTES

1. *Newsweek*, December 28, 1970; *The Wall Street Journal*, January 7, 1971.

2. Effluent charges are more complicated than a simple fee to be paid by each dumper. The fees vary according to how much waste a user puts into a stream and how much contaminants it carries. For a full discussion, see Allen V. Kneese and Blair T. Bower, *Managing Water Quality: Economics, Technology, Institutions* (Johns Hopkins Press for Resources for the Future, Baltimore, 1968). Chapter 12 is a case study of the use of effluent charges by a cooperative of users of the Ruhr River, the best known and often mentioned as a possible model for rivers of the United States.

3. U.S. Bureau of Census, *Census of Governments, 1967*, Vol. 4, No. 5; "Compendium of Government Finances" (Government Printing Office, Washington, D.C., 1969) from Table 7, p. 29.

4. A citizen's easiest access to the federal budget is U.S. Bureau of the Budget, *The Budget in Brief* (Government Printing Office, Washington, D.C., new each year). But read warily. It is propaganda designed to make the President look good.

5. John Steinhart and Marti Mueller, "Search for a Future," Appendix 5, an unpublished report for the Ford Foundation, Feb. 1970.

6. *The Wall Street Journal*, November 10, 1970.

chapter

—— 11 ——

THE PROFESSIONS

» » » » » « « « « « «

DON K. PRICE borrowed from feudalism the concept of estates to describe the roiling mix of science and politics in the United States in the latter twentieth century. There are, he said, four broad functions in all public affairs that reflect the scientific, the professional, the administrative, and the political estates.

These are not sharply distinguished from one another, not even in theory, but they do fall along a gradation of activities. Never are they all one or the other in the minds and acts of their practitioners, for each person does more than one thing and shifts back and forth from one estate to another, but they are useful to define the differences in emphases in our system.

> At one end of the spectrum, pure science is concerned with knowledge and truth; at the other end, pure politics is concerned with power and action. But neither ever exists in pure form. Every person, in his actual work, is concerned to some extent with all four functions.[1]

The professional and administrative estates deal with the knowledge of science, but they add purpose to its use. They also act as interpreters and brokers between the scientific and political estates.

When we contemplate the instruments of social change by which man, and particularly Americans, can save the environment, the professions and administrators are as inescapable as the use of politics and propaganda and the mutuality of government and business. Here we consider the professions and in the next and final chapter, the administrators.

The first discovery about the professions is that there is no agreement about what makes an activity and its practitioners a profession.

DEFINITION

Dictionaries are not much help. Some stress the profession of a faith or an intention; others are so broad that any occupation could be included among the professions. The newest of those consulted, the *American Heritage Dictionary of the English Language,* says a profession requires graduate study in a specific field. This will surprise all those engineers, teachers, dental hygienists, nurses, and medical technicians who entered their careers with only one degree.

When the editors of *Daedalus,* the journal of the American Academy of Arts and Sciences, devoted one of its numbers to the professions, they included articles on law, medicine, the clergy, pre-college education, science, the military, psychiatry, city planning, and politics. When they prepared the number for publication as a book, they added chapters on engineering, architecture, and journalism.[2]

Any of the usually recognized professions will at one time or another be useful in saving the environment. We are concerned, however, with those most consistently involved. They are health, engineering, science, mass communication, art and architecture, and law.

Health. Because the propaganda has been so successful, most people assume that the care of health starts and ends with the practice of medicine by M.D.'s. In fact, many and varied skills protect health, among them nursing, pharmacy, laboratory technology, podiatry, many kinds of research, psychology, dentistry, social work, statistics, sanitation, and all the practices of public health services, including medicine. The chief function of the health profession in managing the environment is to detect and warn when the results of existing practices begin to affect negatively the health of people.

Engineering. Engineers are the chief creators of technology.

They take the findings of science or the ideas of art and apply
them to make machines and to build and design structures. Scientists find that air pollution is caused by the chemicals from automobile and industrial emissions; engineers discover how to reduce the harmful contents. Scientists discover that some organs
can be replaced; engineers design and make the artificial organs
for doctors, also appliers of science, to use as replacements.

Science. The more we know, the more we do. Knowledge
comes from science, or at least that kind of countable, provable
knowledge by which we feel most assured. Other knowledge—
history, aesthetics, philosophy—comes from the broader source of
scholarship, but it is more arguable than the facts of reputable
science. (Selected or biased facts come not from reputable scientists but from scientists acting as propagandists.)

Mass communication. We have not yet sorted out the elements
of the enormous half-business, half-profession by which information and propaganda are distributed in the United States. At one
boundary is advertising designed to sell a particular brand of a
mass-consumed product to as many people as the media can
reach. But closer to the center is advertising in the public interest, copy prepared by the same agencies and space donated by
the same media. Where does that leave a superb television series
on the Public Broadcast network with only a one-line credit to
the business firm that helped pay the bill? Or the professional entertainers who make social comments on sponsored shows?

At the other boundary are the high-brow magazines, the top
newspapers, the best book publishers, and the network television
news shows and news documentaries that refuse to select news
only because it is the most sensational and therefore will draw a
large audience. All these can be misled by the source, but they
will not mislead themselves. They are limited only by the fact
that news is defined everywhere as the reporting of events in
which people are interested. Certainly journalism is a profession
very relevant to telling people what is happening to the environment and what can be done about it. But so is advertising.

Such evasive terms as "communication," or in a debased form
the "communication arts," are no solution to the need for a concept of the mass transmission of facts and ideas. Besides, they

anger journalists of good taste and unnerve professors of English
literature who live in fear of being merged into a department of
radio, television, journalism, speech, and English. "Most working
journalists," says Penn Kimball, "would rather drop dead than
call themselves communicators of anything." And he could quote
the late A. J. Liebling to enforce the point:

> Communication means simply getting any idea across and has
> no intrinsic relation to truth. It is neutral, or the weapon of a po-
> litical knave or the medium of a new religion. . . . Its . . . substi-
> tution, in the schoolman's jargon, for harmless old Journalism dis-
> turbs me. . . .
> Q. What do you do for a living?
> A. I am a communicator.
> Q. What do you communicate? Scarlet fever? [3]

Art and architecture. Practical men of observation and good
sense have recognized for centuries that artists of all kinds—
painters, sculptors, composers, writers, and poets—see what is socially important for the humane welfare of man before those in
the other professions see it. A medical student in his first exposure to the difficulty of defining good health may be comforted
to know that John Donne (1572–1631) saw it long ago.

> There is no health; Physitians say that wee,
> At best, enjoy but a neutralitie.
> And can there bee worse sicknesse, than to know
> That we are never well, nor can be so? [4]

The artists judge by feeling as well as by technique, and feeling includes the contemplation of all the values, including the
value of beauty. Architects combine art and engineering. They
are more important to environment in the artistic side of their
work. In the practice of visual architecture they can do great
good or great damage to man's surroundings.

Law. Lawyers are more than litigious advocates. The good
ones have acquired a way of observing social questions and
thinking about the ways to answer them. The truly well educated lawyers also can write documents that cover the main
points without leaving too many holes for misunderstanding. (If
the purpose is to befog, they can do this too, but the good lawyers know when they are doing one or the other.) And, of course,

when a matter of environment is taken to court, lawyers become the only competent guides through the wonderful world of procedure.

Some will miss education and politics in this list. Education is absent by intention, first because as now conducted it cannot train people to think of the environment in the large sense, but only in segments; and second, because the purpose of education should not be a point of view but rather the introduction of students to the wonder of learning so that they will continue to learn for the rest of their lives.

THE UNCOMMON MAN

Once, before women with grievances made sympathetic the men who are slightly ashamed that the world so revolves around men, Crawford H. Greenewalt could refer in some lectures to the uncommon man. Today a timid editor would recommend uncommon person, in order to appease militant women. Mr. Greenewalt meant the man or woman who rises above the average.

> The story of America is the story of common men who, whatever their motives, whatever their goals, were inspired to uncommon levels of accomplishment. . . . Who can identify the qualities of mind or spirit or dedication in men which mark the division between the common and the uncommon? The strengths and weaknesses of human beings cannot be catalogued as though they referred to blooded dairy stock. . . . The important thing is that we bring into play the full potential of such men, whatever their station.[5]

Mr. Greenewalt at the time he gave the lectures was president of E.I. du Pont de Nemours & Company. He had started with du Pont in 1922, the year he received a Bachelor of Science Degree in Chemical Engineering from the Massachusetts Institute of Technology. Unlike so many others who earned bachelors' degrees that year, he became an uncommon man, not because he was head of one of the great corporations but because he was that and more. He continued to educate himself, adding to his work interests in Roman history, serious music, and, as mentioned earlier, in hummingbirds. He served on boards for other businesses and for various institutions serving a self-chosen duty to society.

At the same time another uncommon man was worried about the same problem of the need for uncommon people in American society. John W. Gardner had stayed at Stanford and the University of California until he received his Ph.D. in 1938. Then he taught psychology at Connecticut College and Mount Holyoke until the Second World War broke out. After the war he went into foundation work as a staff member of the Carnegie Corporation. He rose to become president, and also president of the Carnegie Foundation for the Advancement of Teaching, in 1955. Later he was Secretary of the Department of Health, Education, and Welfare, Chairman of the Urban Coalition, and organizer and head of Common Cause, a new kind of people's lobby in Washington.

While he was president of the Carnegie Corporation and trying to make the best use of its funds for the increase of knowledge and the improvement of the state of man, Mr. Gardner put his thoughts about the uncommon man in a book. He was much too thoughtful for the book to be summarized. One highlight is his recognition that scores made in tests of scholastic aptitude and achievement reveal nothing about the qualities of excellence in a person. Performance in life, rather than in school, emphasizes traits not measured by aptitude and achievement tests, traits such as zeal, judgment, or persistence. Another is his perception that people of excellence have to achieve excellence; it does not just happen to them. They are most likely to want to achieve excellence if the institutions of society expect it or demand it of them and place in their way successive barriers which the uncommon people have to hurdle on their way to achievement.[6]

Both men are dealing with the same point that must be stressed in any discussion of the professions. Both common and uncommon men work side by side in the professions. Only the uncommon men, those who are motivated to strive for excellence, to go beyond the usual performance, to think and act with more initiative, more zeal, better judgment, with more of any of the traits anyone can name, will be much help in saving the environment. The common men will take the easiest road, and, as in the past, the easiest road usually leads to more destruction of the environment. The uncommon men will be the engineer who would rather design technology that does not increase pollution than to accept the design that has been customary, the doctor who drops his professional seclusion and makes public what he knows of the increase in environmental illness and who demands change, the lawyer who seeks the social good rather than the short selfish gain of a client who is wrongfully damaging the environment.

The uncommon men show up only in performance. There are no tests to reveal them and no means to produce them by education. In the 1960's, when money was flowing high, industry and engineering went through a fad-wave of searching for what they called creativity. Lifted out of gobbledygook, the word seemed to mean originality as a personal trait, although the meaning was never certain, for seldom was there such a storm of meaningless talk and frustration as in conferences on creativity. As usual, some academic charlatans were ready to offer Creativity 104, a course in creative engineering and management, no prerequisites, and thus get more attention, not to mention research grants and invitations to conferences on creativity.

The endeavor was futile because creative people are probably uncommon people and they get that way from a whole social sequence and from a desire to be excellent. Above all, they are not to be made from exposure to certain topics in education. As Mr. Greenewalt says:

> You can teach a man the sciences, but you cannot make him a scientist. You can teach him engineering, but you cannot make him an engineer . . . you can teach him executive procedures, but you can't make him an executive.[7]

You watch very carefully men and women at work, and especially the young, and when you see one who appears to have the promise of excellence, you put him, or her, in positions where he or she can be challenged to be excellent. The position may be anywhere, in a business or club, school or church. In the professions, the older members of excellence must be responsible for finding the younger ones, and giving them the opportunities to be excellent. Both the old and young of excellence will have to be patient with ordinary men and women. The common men are seldom willing to change; they do not admit the need; they are too frightened by their own ordinariness to feel comfortable in taking risks on something new.

So it has always been in the United States, where, as Mr. Gardner saw, the three principles of inherited privilege, equalitarianism, and competitive performance are all present.[8]

A law firm or medical staff recruited entirely according to hereditary privilege is doomed, and so is one in which all performance is sweetly equal and excellence is neither admitted nor rewarded. The quickest way to mediocrity for a good university is to treat all professors in rank the same regardless of the quality of their teaching, research, and publications. The merit in "publish or perish" is never recognized by its opponents, who are usually ordinary people. The uncommon professor cannot be happy unless he learns more all the time and tells others about it. He is thinking constantly of facts still unknown, how to find the answers and where to publish the results, and if he is in a good university, he will find the time. Put him in a poor university, where emphasis is not on teaching new and significant information but only on meeting classes, and the uncommon professor decays in bitter unhappiness.

The top flight law firm or medical clinic, and the good university, have enough older members of excellence to spot the young who appear to have the promise. And, of equal necessity, they have escape hatches for the times when they make mistakes: the "trial period for both sides," the temporary attachment, the non-tenured assistant professorship. Those at the top practice competitive performance, double underlined!

WHAT TO DO

The need for uncommon people to lead the effort to save the environment indicates changes in the tactics of selection. Some way is needed to find and encourage all those professionals who can see what needs to be done, who will know best how to do it, and who have the concern and the energy to get it done.

EDUCATION

Men of insight who make a living in the profession of formal education tend to wince when they hear someone say that education is the solution for any social problem from a high birth rate to street litter. It cannot be more than a part of a solution and then it will do not what the advocate has in mind, which is propaganda, but what educators conceive to be their mission, which is to teach skills, facts, theories in a broad offering of what has been defined as the main subdivisions of subject matter. Education deals with language, art, mathematics, natural science, social science, history, and the skills of vocation. Only when the morality of social problems comes up in courses in these subjects can formal education be the answer to a question of purpose.

Yet at one time or another every thinking professional man, bothered by the prevalence of tunnel vision, will become interested in the curriculum as a means of broadening awareness. He will wonder if doctors should not have more courses in sociology and psychology, if engineers should not be exposed to political science and landscape architecture. His next turn of mind tells him that his own profession has become so demanding of skill that it is a wonder any student is able to master enough of it to get started in the work. Engineering is chemical, civil, electrical, mechanical, metallurgical, mining, or nuclear. Law is the law of contracts, procedure, torts, property, trusts and estates, corporations, taxation, administration, labor relations, evidence, and crime. Medicine is general medicine, but it is also anesthesiology, oncology (cancer), gynecology and obstetrics, neurology, psychiatry, pediatrics, radiology, surgery, and prevention. Underlying all doctors' skills are physiology, chemistry, pharmacology, anatomy, genetics, and pathology.

Within each sub-section of all the professions, the knowledge and elementary skills to be mastered have grown at the now familiar exponential rate. One solution, and the one reached in the more advanced and socially conscious professional schools, is to remove most if not all the requirements for pre-professional courses. These schools encourage broadening by saying, for example, that for admission to medical school an applicant must show only a small number of hours in chemistry, physics, and zoology, and most of his pre-medical education can be, some schools say should be, in subjects not offered in the medical school. A full four years of general college is recommended.

Another solution is to extend the time required to get the professional degree and to place courses of social concern in the professional curriculum, where they can be related to real questions that must be answered. Only a school of outstanding reputation considers this one very long. Because costs are high and students want to get to work, enrollment might drop in the average school unless all schools made the change. Competition in the academic world makes a simultaneous reform for the sake of more social awareness only a possibility.

WHAT TO ENCOURAGE

It would be foolish to say that education makes little difference in making the common man uncommon. Schools and the mass media are the only institutions of society that reach every American. Schools take all Americans for a few years and a good percentage of Americans until they leave college. Education, at the same time, can be counted on too heavily.

The mistake comes from assuming that all students exposed to the same subject will take it to their hearts and minds and never let it go. So we are told that all students in public schools should be required to take courses in conservation, hygiene, Americanism, Afro-American history, physical fitness, or whatever the promoter of the moment advocates. Once tutored in these subjects the citizen through life will conserve resources, be healthy, and be a good American who respects the dignity of black Americans, who now have more dignity because they have been recognized in a history course. To deny this from experience in the classroom does little to convince these advocates that they talk nonsense. Their cause is always the holy cause.

That some Americans come out of required courses in English composition unable to write, spell, or think, that they come out of any course only with the interest they took into it, is hard to explain. That the content of college courses, especially in the best colleges, varies according to the interest and knowledge of the professor is a trade secret kept from people off-campus and sometimes from colleagues on campus. And what happens to students in all grades and all courses depends on the quality of the teachers.

Despite these truths some practices should be encouraged to allow individuals to grow as much as possible, whether in school or at work in the professions.

ALLOW DISCUSSION

While by definition a professional man will spend most of his time practicing skills, there is no reason why he should not be allowed periodically in the course of his study or work to join others of similar interest to discuss the social aspects of his profession. The uncommon men thus may discover one another; the common man who attends from curiosity might discover that he is uncommon but did not know it. One of the surprising revelations to come out of the great explosion of interest in the environment was how many scientists and other professional people were concerned but until then had not had a movement or cause in which they could work. Students and faculties alike in the natural sciences and professional schools discovered that as citizens they could do something and that many others like themselves were concerned. Why not provide two hours each month in professional courses to discuss the profession in its social context and the environment or whatever else is important to society? Why not provide professional conferences periodically for similar discussions among those who want to talk? The meeting can be face to face though it need not be if travel is too much trouble. A conference can be conducted by telephone or closed circuit television. A speaker can be on tape if he cannot attend in person. The important thing is not a speaker but discussion among those members of the profession who are concerned. Professionals spend a great deal of time keeping up with new advances in their fields. Some might like to keep up with changes in the consequences of their work and their responsibility to society.

ALLOW INDIVIDUAL STUDY

Despite all the talk about the inaccessibility of professors, the typical college student seldom wants to be faced in a private confrontation with a professor. His aim is to get out of college as soon as possible. He can fake most of his work and still make the grades he aims to make.

A elaborate ritual of self-deception has obscured what really happens in colleges until most of the participants think that what they only believe is truth. One belief passed on among students is that professors should spend more time with students. A belief passed on among professors is that courses and syllabi should be designed to cover as much as possible of the subject, more than students need for an introduction to learning.

One result of the preoccupation with courses is that the so-called "reading course," or tutorial, or honors course, senior thesis, or whatever it is called, is used less than it could be for the encouragement of uncommon students. Lazy students try to use it in order to avoid work. Wise professors refuse their requests. Exploitative and selfish professors try to use it to get free help in exchange for credits. A wise student refuses to work on a subject he does not choose for himself.

People with good intentions can use the reading course to enter a subject of special significance, about which neither student nor professor knows as much as he would like to know. The student can try out his interest, learn how to use new sources of knowledge. He can grow faster. The professor can test a student he thinks might be uncommon to see whether he is or not.

When the reading course is not available, uncommon people in schools and colleges can get the same results but without the credits if they will only take the time. All the uncommon professionals need time to think. It is much easier to find time in school than it is later on the job.

THE TALK MISSION

No one who is not on the invitation lists can imagine the enormous capacity of organized Americans for listening to talks. They listen at luncheon clubs, at high school assemblies and club meetings, at church meetings, trade and professional meetings, at ceremonial banquets and breakfast prayer meetings; wherever two or more are gathered they seem to appoint a program chairman who proceeds to get speakers for any time or any occasion. The uncommon man in any profession is a good prospect for the list of possible speakers. If he does a good job the first time, he will be invited to other groups. He must then decide whether this role is a good use of his time and affords him satisfaction. Lack of time is the great problem faced by uncommon men. They must live by schedules so tight that almost nothing can be random. If they decide to accept invitations to talk before groups, they must sacrifice some other use of their time.

Such talk is important for any subject, and thus for questions of environment. It gives the professional who knows most about his subject an audience, most of whom will be listening and hearing as accurately as people can hear. He can persuade through knowledge, if that is his purpose. And he can, by his presence and manner, leave the impression that special knowledge is a good thing to use in the solution of such problems of society as the use of environment. When so much thinking about social questions is based on emotions or selfish interest, the display of knowledge is a good counter experience for any organized group.

A SOCIAL AUDIT?

Proposals for social audits have become fashionable recommendations. They include citizens' councils to watch the police, neighborhood councils to advise administrators, youth councils to advise public officials, public groups to keep an eye on the professions that most affect the public interest. Prisons, universities, and some hospitals have had experience with Boards of Visitors.

The experienced public official or professional man, especially the uncommon men among them, know that the competence of such boards depends on their political strength in the organization and on the way each member of the board defines his personal role, or whether he is a friend and protector of those watched or their vigilant critic. If a board of visitors inspects only once or twice a year, it seldom knows enough to be heard respectfully. If it deals only with officials who do not have the power to make decisions, it will be deceived at worst and shown selected information at best. If it has no way to report to a clientele of power, it wastes its time because it cannot bring enough pressure to force changes.

All these defects are illustrated by the board of visitors for a university. The board visits the campus rarely and never stays long enough to sense the real troubles. It deals with administrators and sometimes meets with trustees, whereas the centers of power for academic matters are the departmental faculties and academic matters are the heart of the quality of education. It has no clientele of power. It has no supporting group at all unless in some rare time, perhaps, the alumni association generates enough pressure to force the faculty to listen.

A university faculty in a system of strong departments is about as independent as any profession on earth. Within each department the members can claim that they know more than any group what is important in their subject. When a change is suggested in curriculum, they can simply ignore it. If they feel generous, they can explain patiently, once only, why the outsider is wrong. In this system the members of another department within the same institution are outsiders. Members of one department do not make suggestions to another department but all will support a department whose autonomy is questioned. In a good university professors are the ultimate free men.

Professors illustrate the difficulty of any social audit of the professions. The real trouble with education is not access to professors or any of the demands that agitators or militants have spotlighted. It is the curriculum. The professionals in each department have let it accumulate. Some exotic, minor courses are added occasionally to please the professors who know the subjects. Others, especially in the departments that teach skills, are added because committees on professional education in the departments' clienteles requested them. Some, alas, were added because a foundation or a government offered money to teach the subjects. And in the trashy departments some were added for public relations by professors who wanted to attract large enrollments.

Yet how can any outsiders crack this insularity? Who knows more about what should be taught in chemistry than the members of the chemistry department? Who knows more about English literature than the members of the English department? Or about sociology, art, political science, physics, or any other subject? True the curriculum wastes students' time, and too many courses often are irrelevant to student needs. But an outside

group that starts out to design a new curriculum for any department or for a whole college starts with a steep uphill road to
climb, threatened from all sides by the specialists.

The answer in terms of social implications, in this case in
terms of sound management of the environment, is not in social
audits. Change will occur only when the uncommon men in a
profession band together and persuade enough of their fellow
members that change is needed. Change of this kind takes place
steadily in the professions, although it is usually slow. Generations of lawyers and of persons accused of crime may come and
go while other lawyers argue over the needed reform of criminal
justice. A geological sense of pace may bring change too late to
save justice or the environment.

At the same time the change of attitudes and practices in the
professions concerning their role in saving the environment must
come, in our present method, from the professions. Only brute
force can impose change from without. And history shows that
any time the professions are subjected to brute force, the quality
of their work and talent declines until all society is damaged by
unequal justice, distorted truths, and misconceived values.

NOTES

1. Don K. Price, *The Scientific Estate* (The Belknap Press of Harvard
University Press, Cambridge, 1965), p. 135. This is one of the few
original books. For those who missed them in the history books, the
feudal estates were the nobility, the clergy, and the commons. A sov-
ereign would occasionally consult the three estates.

2. Kenneth S. Lynn and the editors of *Daedalus* (eds.), *The Professions
in America* (Houghton Mifflin, Boston, 1965).

3. Penn Kimball, "Journalism: Art, Craft or Profession?" in Kenneth
S. Lynn and the editors of *Daedalus, ibid.,* pp. 244–45. Liebling is
quoted from A.J. Liebling, *The Press* (Ballantine Books, New York,
1961), no page number given.

4. From "An Anatomie of the World—The First Anniversary," *The
Complete Poetry and Selected Prose of John Donne* (Modern Library, New York, 1969), p. 188.

5. Crawford H. Greenewalt, *The Uncommon Man, the Individual and the Organization* (McGraw Hill, New York, 1959), pp. 1–3.
6. John W. Gardner, *Excellence* (Harper & Row, New York, 1961; also Harper Colophon edition, 1962).
7. Greenewalt, *op. cit.*, p. 76.
8. Gardner, *op.cit* (Harper Colophon edition), p. 21.

chapter

—— 12 ——

BUREAUCRACY

»»»»»»«««««

BUREAUCRACY is essential in any advanced society. Only a primitive society can get along without it. Once the work gets organized, once people are assigned what to do, once other people begin to coordinate the various distribution of tasks, the administrative process has been introduced and people have begun to work in an organization. This is all the term "bureaucracy" means to a lexicographer: people working in units on assigned tasks. In recent times usage has been limited to government, although any so-called private organization will also exhibit the same condition. There also has been a tendency to limit the term to the executive function of government, although courts and legislatures also have executive functions.

Bureaucracy is one of the least understood and most berated words in man's vocabulary. Academic writing about bureaucracy has a ghostly unreal quality for the reader who knows the subject from experience. It recognizes all that has been learned about human behavior in organizations since 1930, but it cannot shed its deference to Max Weber, a German sociologist who dreamed up a theoretical definition of bureaucracy and who is almost always quoted in the first footnote.

A bureaucracy is a society within society, and it exhibits all the same traits of human behavior. The most recent and accurate social science research has proved this. Experience inside a bureaucracy proves some other facts, that some day perhaps also will be discovered by the social scientists.

One of the credos in most of the academic literature is that rules guide all behavior within a bureaucracy. Those with experience, who are adept at doing what they think ought to be done, know that rules can be changed, or by-passed, and often one rule can be used to negate another rule. Sometimes the most successful, most esteemed, director of administration, the official who has most to do with the rules concerning money, personnel, services, and facilities, spends most of his time finding his way legitimately through the rules to accomplish what he wants done. He does not so much enforce rules as use them.

Another credo holds that because bureaucracy can assert control over the lives of citizens, it has a tendency to do so and is a threat to freedom. Actually, in America, the bureaucracy is composed of people holding the same attitudes as citizens outside it. Until the culture changes, the bureaucracy is but a small threat in itself. If anything, it is more timid than aggressive. Whether some part of the bureaucracy, say the professional military or a big city police force, would continue to be benign when their work is threatened or when a political leader decides to use them in socially unacceptable ways is a question not yet raised beyond a vague worry.

Still another credo holds that bureaucracy is organized by hierarchy, and this is true in the sense that some people are above other people in rank. But the corollary that a superior can issue an order and be certain that it will be carried out has been disproved by both research and experience. Even the President's orders can be ignored unless he checks more than once. Jonathan Daniels, an aide to Franklin D. Roosevelt, once observed:

> None [of a President's immediate subordinates] are insignificant. Their jobs are news; also, all of them maintain information men to keep them in the news. . . .

> Half of a President's suggestions, which theoretically carry the weight of orders, can be safely forgotten by a Cabinet member. And if the President asks about a suggestion a second time, he can be told that it is being investigated. If he asks a third time, a wise Cabinet officer will give him at least part of what he suggests. But only occasionally, except about the most important matters, do Presidents ever get around to asking three times.[1]

What is true for a President is also true for the head of any large agency, and practically all of them in the federal bureaucracy are large.

Finally—and this can be the last credo for there is little point in raising others except to show misunderstanding—another credo states that bureaucratic organization of itself produces efficiency. Any observant practitioner will quietly laugh at this one. He will have seen work created to keep people busy so they will not have to be fired. He will have seen men try to get on committees because, for them, attendance at meetings has become a career. He will remember times when the only person who accomplished any new purpose was the individual who found holes in the vast structure and who wiggled through them undetected.

A cliché prominent in the late 1960's was that bureaucracy in any form was a threat to individual freedom. It supplied rules and enforcement when the ideal society should allow the individual to choose for himself. It represented the views of the "establishment." To these accusations, the establishment could only agree. This was certainly the function of the bureaucracy. It expressed the views of the overwhelming majority of society. It was composed of the same kind of people as the majority in the society around it.

To a political scientist concerned with saving the environment, the great significance of the bureaucracy is that it is useful in the cause. Individuals within the bureaucracy are in as strategic positions as are businessmen, legislators, lawyers, and other professional people, and they have as much authority to initiate and to manage the job of environmental salvation.

Because the federal government will be the main rulemaker for the environment, its bureaucracy is the one that matters most. It has certain traits observed from experience, though not covered in the academic literature.

A FEW MAKE THE DIFFERENCE

Just as a few manage the affairs of society through politics and propaganda, so a few also set the tone, make the decisions, manage the affairs of the federal bureaucracy. They are the ones who take the trouble to excel. They are very much like the outstand-

ing people in any group. Work is more interesting to them than
watching the clock. They take sick leave only when they are sick,
when the average federal workers make sure that they collect all
the sick leave owed them. They ignore coffee breaks. They work
at lunch. At any rank of the hierarchy, they strive to do the best
possible job, not just the acceptable job.

One of the most disturbing ironies of life is that a few are motivated to excel while most are satisfied to be average. The conditions that motivate the few appear not to be the same in all
cases. No solution can be found, no prescription written. One
person excels because he admires his parents, another because he
vows not to be like his parents. Peers can sometimes inspire, but
usually peers, who are in the majority average people, will re-
ward conformity; and the saddest event in life is to see the gifted
high school student wasted by the pressure to be popular. Fads
of speech, fads of music, and distaste for learning crowd out the
promise of creation and excellence. The youth who excels will
most likely be lonely in the crowd until he enters a career and
finds others of like mind.

The familiar drawbacks of bureaucracy such as inertia, make-
work, picayune adherence to minor rules, are practiced by the ordinary men and women who do not realize that they are a waste
of time or who realize their pettiness yet need it as a way to
make a living and to express their personalities. Those who excel
do not have time for them.

In the federal bureaucracy the few who excel soon get to know
one another. They identify by sharing a viewpoint. Liberals, con-
servatives, realists, sentimentalists, ruralists, urbanists, wilderness
advocates, Kennedy people and Nixon people, flunkies for an
agency's clientele—any viewpoint will have believers who stand
out from the average. All those of the same viewpoint within a
group are the good guys; opponents from other groups are the
bad guys. Administration takes place in the competition of such
groups. The characteristics of bureaucracy are the products of
this competition.

On the whole the more generous have won over the more selfish in the federal bureaucracy, and the people of breadth over
the people of narrow purpose. The leaders in ideas have been

people with a concern for the general welfare rather than the particular welfare of one business firm or a few individuals. Of course, sometimes they see the general welfare as being served by benefit to a few, but, not so odd, this can be true since we are a nation of groups and the federal executive is organized to serve such particular interests as veterans, labor, farmers, business, users of atomic energy, common carriers, scientists, artists, and nearly any other interest in the catalog.

Among the excellent who have so far dominated the federal bureaucracy, a profession of public service has formed. It includes the political in-and-outer who works while his party's President is in office. Not much noticed, a group of excellent business executives, lawyers, and professors in this nation choose to be in private work through one President's administration and to be in government in the next. The profession also includes permanent civil servants, some of whom will be appointed by a President to be political aides during his term. The old firm line between the political and the permanent public servant has all but disappeared in Washington among those who excel.

The dominant members of this profession, under a President of either party, share the same purpose—to make the nation a less troubled place, to keep it strong, to serve the clients of their agencies, and to work for the President's program. They have advantages over citizens outside the bureaucracy. For one, they have access to enormous amounts of open and secret information about a great variety of things. Much of it is in the files collected by the bureaucracy itself. If not there, it can be gathered by the worldwide system of federal offices.

Information indicates where the new problems will be and what can be done about them. This is followed by initiative: a new program or change in one existing, a new bill to be introduced in Congress, a new emphasis. Nearly all legislation acted on seriously by Congress, and not introduced just to please some constituent, arises in the bureaucracy.

The few who make things happen soon get to know some members of Congress. They know the influential members of the sub-committees that deal with their agencies; they know when the strategic decisions of Congress are made in sub-committees and committees. An individual in Congress needs something for a constituent; he calls his friend in an executive agency. His staff gets to know people in the agency.

Everyone works for the same employer, the citizens, and any arguments are not about loyalties but about tactics and priorities. Executive officials have a special position as lobbyists. They can deal with members of Congress in a way that no representative of a profit-making group can deal. And they do so all the time.

Another advantage is in propaganda. By custom, news media will cover executive agencies, sometimes as routine and sometimes only when called. Government works for the public and not for private gain for its officials. Reporters will watch business firms like snakes to make sure they do not get free some publicity that they should pay for as advertising, although this varies too and the sports and business sections of newspapers give many free rides to profit-seekers. Newsmen do not act as if a government official is after a free ride. When he promotes a program, he does so for the public, in theory.

BUREAUCRATS CHOOSE PROGRAMS

The federal bureaucracy is not a neutral and passive body which responds to all requests, pressures, demands from either citizens or members of Congress. It makes its own requests, brings its own pressures, makes its own propaganda as its leaders decide. This fact that leaders among the bureaucrats decide what to advocate and what to oppose creates perhaps more misunderstanding among inexperienced laymen than any other trait of government. The laymen cannot understand why bureaucrats simply will not advocate some action that makes such clear good sense to them that no one should ignore it.

Two considerations are always in the minds of the leaders. One is the inevitability of precedent. The other is the constant necessity to make choices in the budgeting of time and money. Whatever a government agency does for one citizen, it must do for all citizens who fall within the category that receives that service. This is a fundamental principle of equality. It is true for regulation as well as service. The government of a society of legally equal people must treat all alike within the group to which a law applies. This means that any new action becomes a precedent to which the agency must conform in all future cases unless, unlikely, Congress should take away the authority. A wise bureaucrat watches with care every innovation to see what it will mean when extended to all who will be entitled to the service or affected by the rule. If he does not want to spend his agency's time and money on the precedent expanded, he will not make the innovation.

The budget is the other concern, and a big one. A leader of the bureaucracy spends much of his time deciding how much to request for which purposes, then pleads his case before the budget officers in his agency, his department, and the budget officers who represent the President. After he has worked his request through these control points within the executive branch, trimming usually at each point, he must argue his case before the subcommittees of the committees on appropriations in the House and Senate. There is never enough money to do all that the official knows should be done.

The official must think also of another kind of budget. With the money, time, and staff available, what kinds of emphases shall he follow? The choice will mean that some laymen will think that the bureaucracy is not doing some of the things that should be obvious to anyone must be done. The laymen think it is either blind or obstinate. It is neither. It has chosen other emphases and does not want to spend the time or money for what the laymen argue for.

Federal executives and the influential members of their staffs decide what to propose and what to emphasize. Congress by enacting a law gives the original assignment and adds to it as years pass. The executives usually suggest the changes. The ideas can come from within the bureaucracy, from a special interest group, or from an individual member of Congress. Always within the enabling law, the executives will make decisions that are not spelled out in the law. These will deal with how to organize and how to manage, which parts of the task to do first and which to postpone, which to push and which to give only token notice. From all these initiatives within the executive agencies, the few who make the decisions produce the complicated nature of federal administration and relations with Congress.

THE BUREAUCRACY IS POLITICAL

Some members of the federal bureaucracy are partisan. They enter with the election of a President and form what is called his administration. At the top they are secretaries (heads) of departments, heads of agencies not called departments, members of boards and commissions. In the lower ranks they are aides to the partisans at the top and to the President.

More of the bureaucrats are permanent, employed by some kind of examination or by a test of qualifications and holding their jobs so long as there is money to pay them. Nearly all the ordinary rank-and-file people who do not strive for excellence or who do not care much about the purposes of their work are in the permanent civil service.

Politics in the bureaucracy has three shades of meaning. One is the effort of a party to hold support and gain votes for its President or candidate for the presidency. Another is the contest to get a policy adopted in competition with other policies that are being considered. The third meaning is the contest to make a policy successful after it is adopted.

Members of the President's administration are responsible for keeping his name and his party's name in good favor. Both partisan and permanent officials of all ages and ranks get involved in the other two kinds of politics. As usual, only the few who care enough to make the extra effort will be the ones who handle the politics of bureaucracy.

For them life is filled with plans, coordination, debate, negotiation, and hard work. They are specialists at inter-personal relations and getting agreement among a number of people. They know how to talk and write with clarity. They can face hostility without taking it personally. They know how to think clearly in tense encounters.

They live in competition, for that is what the practice of politics is. Just as there is never enough money, time, or staff to do everything that ought to be done, neither is there unanimous support for one policy as against another. Agencies compete with one another for one policy or another, for one provision or another. People compete with one another for policies and for power. The fight over jurisdiction is standard in the federal bureaucracy. Federal agencies fight one another. They also fight state and city agencies, and the arguments are reciprocated. Interest groups fight one another often with allies among the competing bureaucrats. Interest groups argue with bureaucrats over proposals that will affect their members. All the arguments are about what the various participants think will best accomplish a purpose or protect an interest.

All the competition of a pluralistic society is found inside the bureaucracy and in the relations of the bureaucracy with Congress and interest groups. The resolution of competition is the way decisions about the environment will be reached. Almost never will the advocates of any proposal have an easy time or get what they want all at once. Change in America comes by steps and over a span of time. That is the way we do things.

The federal bureaucracy is political in another way that can threaten its ability to be responsible to the general welfare. It is always a latent target for the propaganda of reckless propagandists. These selfish and irresponsible enemies of order and progress toward the good nation and world have been busy since the early 1930's in their contemporary phase. They conduct investigations, with prejudice, by one congressional committee or another, usually for signs of what they call subversion. They are a constant threat to the reputation and career of any concerned member of the bureaucracy.

When the nation is worried, when the mass media lose their sense of fairness, when a linking of events draws attention to the attacks, the enemies of good government can have their day, as they had from the time of President Truman's loyalty boards of the late 1940's until the downfall of Senator Joseph McCarthy. During such a time, a federal bureaucrat can only lie quiet and hope that he will not be framed in the mass media on false evidence. He will not take initiative. He may resign and find other work. He is denied his duty to say what he thinks will be the next desirable change in the national interest. The bureaucrats can be their most useful to society only when the great majority of citizens in that society trust them to be loyal, able, and thinking more of the public interest than of their own.

BUREAUCRACY IS BIG

A human can no more comprehend the bigness of the federal bureaucracy than he can comprehend the size of the national budget or the number of miles to Saturn. There are between 5 and 6 million military and civil executive employees, fewer than the populations of Tokyo, New York, London, or Mexico City, but more than Chicago or Philadelphia. Always 10 percent or less work in Washington. The rest work in federal offices in the states and in other nations. All of them have never been gathered together in one place so that an idea of their numbers might be grasped.

A chart to show the whole organization would occupy acres of space and take years to prepare. The nearest approach to a statement about all the sub-divisions is the telephone directory for a department, in the section organized by units. The telephone directory shows only the units, in Washington or some other location.

Men and women in all the units in Washington, in the states, in the world will make decisions that affect the environment. The few who manage the affairs of the bureaucracy will be located in all the units; a park ranger in the Big Bend may be able to do more than the Secretary of the Interior because he is closer to the earth and less tied up in conferences and clearances.

Several facts of life are created by the very bigness of the federal bureaucracy. The superiors shown on a chart cannot possibly know all that goes on in the units they theoretically command. Olympians may appear to be in charge, and they talk as if they are in press conferences, for which they have been briefed by public relations men whose job is to anticipate the questions that reporters will ask. In fact, a boss on Olympus knows only what comes to his attention. His subordinates make the choice of what to bring to his attention. They do not find it easy to reach him. He is busy with his own peers, with public relations, with the politics of legislative relations.

The vastness of organization and the inability of Olympians to supervise details means that this bureaucracy is full of holes through which men of skill and self-confidence can do many things not specified in law or executive orders. As long as the act is within the broad definition of the law, the act will not be noticed widely and the actor will not be reprehended. Only the ordinary men and women look on bigness as an excuse for doing nothing. The able ones use the freedom created by bigness. As usual, they can use the freedom of bigness to save the environment or to damage it, and they use it both ways.

In the units that are remote from Olympus the work gets very specialized. The rules for use of a particular pesticide will be decided, and recommended for approval up the line, by one or two chemists who are interested first in the chemistry and the economics of their substance and second in what it might do to the environment. They will see more of chemists from the industry making the pesticide than of superiors with a broader concern.

Because laymen concerned with the environment usually do not know how big bureaucracy works, and spend time with *ad hoc* commissions, with agency heads, or with congressional committees, the pesticide chemists will know about the crisis of the environment only by what he gets from the press. If he gets worried, he will speak to his friends, the chemists from industry who work on the same pesticide, and together they may make some changes in their product and the rules for its use. If he is not concerned, he can continue his way unnoticed for years. He may never be brought to account unless he damages the environment so much that it becomes news.

Bigness is also responsible for the truth of all statements that bureaucracy keeps rolling from momentum after individuals have seen the need for change and that change comes slowly. Momentum may account for most of the inaction that hurts society. Bureaucrats get set in a course and continue in it after change is indicated. Generals who were young field grade officers at the end of the Second World War become directors of strategy in a guerrilla war. Their habits, firm in the momentum of a huge bureaucratic apparatus, lead them to think they can win by the same old use of superior firepower. They unload tons of bombs and artillery shells to stop a few men with rifles, machine guns, and a mortar. The enemy simply moves out of the way. No fronts can be drawn nor cities captured. The enemy is fluid.

Highway builders devise a scheme of roads planned around dollar economies and cannot change easily to another system of planning that would consider the quality of life. Yet the new roads destroy neighborhoods and quiet urban vistas.

The bigness of bureaucracy means that the individual becomes more important than ever. When so many decisions that affect environment are made at a distance from Olympus by people who are unknown except in the small fraternity of the like-minded, the individuals who decide are the key to environmental salvation. We are brought again to the fact that social change, in politics and propaganda or in bureaucracy, starts with an individual.

HOW TO WIN IN BUREAUCRACY

None of the textbook rules about the responsibility of the bureaucracy pretends to insure that individuals will always do the right thing about environment or any other concern of society. The tactics for getting successful change through the use of the federal bureaucracy are derived from experience. They are not written down but are carried around in the intuition of all who practice successfully. They are used by the professional lobbyists and by the successful bureaucrats, and when a member of Congress or a member of his staff wants something done in a far remote corner of the organization chart, he will act the same way.

The first rule deals with purpose, and the first point sounds so simple that it can be faulted until one learns how often it is violated. It is to know your purpose. Know what action you want. Be specific; be clear. If you want to clean up Possum Creek, don't talk about water pollution in the abstract. Talk about Possum Creek and its idyllic past before industry came and before urban and rural solid waste begat its ugliness. Bureaucracy is organized not to work on generalities but on one thing at a time. Bureaucrats do not operate on the whole environment nor even water pollution as a whole. They act on one Possum Creek after another, one stretch of river, one polluting industry or one polluting city at a time.

Congress deals with the general subject. It adopts a law authorizing a particular unit of the bureaucracy to take certain actions under the law. Men and women, individuals in that unit, take the actions, interpreting the law as they go along. They take action one case at a time. If a citizen goes to court, pleading either to stop an action or to make the bureaucracy take action, the court too will act only on one case at a time.

The second tactic for success sounds equally simple, but it must be said because it is so often ignored. It is that federal bureaucrats with only rarest exception are meticulously honest and honorable. The assumption among the ignorant that government can be bought will lead to failure if put into practice. Not since the Harding Administration has there been a national scandal that involved executive officials. Under Harding, it was two top members of an administration that also included some of the most honorable men in the land—Henry C. Wallace, the father of the later Vice President Henry A. Wallace, Herbert Hoover, and Charles Evans Hughes, to name three.

The more recent case of Sherman Adams, who was top aide to President Eisenhower, was more a tragedy than a scandal. He had accepted gifts, an overcoat and payment of hotel bills, from a friend for whom he intervened with one of the agencies. No doubt he considered the favor he did as wholly separate from the gifts. He was a strict man. He was caught in what looked bad for one so close to the President.

The rule is, then, never ask a federal bureaucrat to do anything that is unethical. His profession of public service has an ethic as firm as that of the law or medicine. If he is asked to violate it, he will refuse, and he will have been insulted.

The other rules for success in working with the bureaucracy are less apparent if we judge from listening to laymen discuss politics or from dealing with laymen from inside the bureaucracy. Get acquainted face to face with the others who share your purpose and who are involved in the same situation. You now have allies in a group, when government by groups and decision by groups is so prevalent in America. You can help and encourage one another. Nothing is hopeless so long as several good men and true are united, and because you will consider your purpose a good one, you will see yourself and your associates as good.

By definition, your opponents will not be good. In fact, they merely have a different idea of what purpose should be served, and inside their group they think your group is wrong. It matters little what you think of them—in the next cause they may be your allies—the main thing for tactics is that for now, and your present purpose, they are your opponents in the competition of politics. Among the successful, the game is fair but not generous.

You will have secrets that you guard with skill. You may openly get opponents involved in the pursuit of some other purpose which is also good, but your tactic is mainly to keep them busy while you work around them. You may offer a compromise which has less meaning than it seems to have. In this game every man is expected to be smart enough to play, and if he fails, he must accept the fault. The smart men, therefore, never take any statement, act, or proposal at full face value. They are always alert to the true meaning. But they do not lie or break their word. Secrets are permitted in politics, but not lies and false promises. These brand their executors as men not to be trusted or respected.

Know where the power is and go straight to this point. It will differ for each issue. Amateurs frequently think that if they go to the head of a department, they are at the center of power. They waste their time, for the weakness that is true of the President is also true of the head of a department. He can make a suggestion and unless his staff keeps it alive, his suggestion can be ignored.

Often the subordinate who writes the first draft of a policy in a sub-sub-unit has more influence than all the others who get involved at later stages. The first draft sets the agenda of what will be considered. If a later participant thinks a different proposition should be considered, the chances are he will not have time to write a new paper, and if he does the other participants are already preoccupied with that first draft and subconsciously defending it because they have already become involved with it. Titles on the office door have little to do with real power in a bureaucracy. The power of leadership is created by the situation for each decision.

Beware of myths and wasteful preoccupations, and do not ·
waste time dealing with the drones. The most successful of those
who excel learn that much of what ordinary people believe is not
true but only what they have been conditioned to think. In business the ordinary people think that government and business are
really separate and that business stands on its own as private enterprise. The excellent know this is a myth, and they pay as
much attention to the policies of government as to business because what government does affects business.

When one of the excellent, Robert Townsend, wrote a horn-
book that cut through many of the business myths in alphabeti-
cal order, it became a best-seller, although it was a puzzle to
know what the ordinary men of business would find in it save
annoyance. He summed up the danger of myths in one obvious
but heretical statement:

> When the vast majority of big companies are in agreement on
> some practice or policy, you can be fairly certain that it's out of
> date. Ask yourself: "What's the opposite of this conventional wis-
> dom?" And then work back to what makes sense.[2]

In the federal bureaucracy the ordinary people believe that decisions really are made through the strict channels of the organization chart. The excellent get together, reach agreement, and
then fix up the papers to go through channels for approval.

The excellent also question all new practices to see if they
really are improvements, for they have seen some of their own
kind defeated by falling for some false promise. Fads are always
offered along with the truly better way of acting.

Computers are a great improvement for certain kinds of work.
Some good men made the mistake of turning to computers without first (a) making sure that they would be more efficient, and
(b) making sure that the information that went into them was
accurate. The computer will gladly produce tons of paper devoted to information that is not really needed but which is produced because blind managers were told that it could be had.
The handling of paper can smother other work. A computer is
only an elaborate office machine. It does not create fact; it only
processes data put into it by humans who can err. To return to
Mr. Townsend's horn-book:

First, get it through your head that computers are big, expensive, fast, dumb adding-machine-typewriters. Then realize that most of the computer technicians that you're likely to meet or hire are complicators, not simplifiers. They're trying to make it look tough. Not easy. They're building a mystique, a priesthood, their own mumbo-jumbo ritual to keep you from knowing what they—and you—are doing.[3]

Robert McNamara, the Secretary of Defense for Presidents Kennedy and Johnson, started out to be the most successful of latter-day leaders of the federal bureaucracy. He brought the military establishment under civilian control for the first time, imposed a coordinated budget on the four military services and thus made the navy, army, air force, and marines begin to think alike, and he began to move against nuclear testing and for some kind of assurance that no nation would ever use nuclear weapons.

Yet Mr. McNamara also presided over the enlargement of the war in Viet Nam, and the choices of strategy in fighting it, until he, like President Johnson, was held responsible for what had become a national disaster. David Halberstam thinks McNamara's blindness came from knowing too much that was untrue.

McNamara, who *knew data,* would go over it more carefully than the military. Hence the portrait of McNamara at his desk, on planes, in Saigon, poring over page after page of details about each province, each district, each company, battalion, platoon, squad. All those statistics. All lies. . . .
    He studied a guerrilla war which always seemed to quantify so well—a feudal army backed up by the enormous firepower of the most powerful nation in the world, with tanks, airplanes, and helicopters, and fighting an essentially conventional war, will always have better kill statistics than a modern peasant army which uses its limited power in the refinements of guerrilla war. Add to that the fact that all Vietnamese commanders were liars. The reports they sent in were all lies since they never dared admit that they might lose one battle or suffer heavy casualties. If the American advisers knew that these were lies, they soon found out that MACV (Military Assistance Command in Vietnam) did not want an open challenge to the reporting. . . .
    McNamara arrived time after time for his on-the-spot visits, always acting out the carefully charted tours set up first by General Harkins and then by General Westmoreland. . . .

What was created on those trips was not a knowledge of the country, but something worse and far more dangerous—an illusion of knowledge. He was getting the same information which was presented in Washington, but now it was presented much more effectively in Vietnam. McNamara was not cynical; he did not know any better.[4]

A computer's output is only as good as its input, and, to continue in the jargon of the trade, there is great truth in GIGO.[5]

A drone is a male bee that is stingless, does no work, and produces no honey. In its human version it does some work. One of the myths of bureaucracy is that the work done by drones is an important part of administration. In fact it is a by-product, made up mostly of paper to be written, carried, and filed, and as often as not unnecessary but continued because no one is mean enough to fire drones and cut expenses. A drone can be found anywhere on the hierarchical ladder, although seldom near the top, where he would be exposed. The people who manage the affairs of bureaucracy do not waste time with drones. They have more important things to do.

Some other cautions can be listed for those who would succeed in getting change through bureaucracy. In certain situations, you should let all the public credit for accomplishment go to some other person. You will have to know when this is true. Generally speaking, you always let the credit go to a politician who has to run for re-election. He needs the publicity. If you are inside in a staff position, usually the credit should go to your executive. The people who matter will know who really accomplished the change, and you don't need the publicity. It is cheap in government.

Do not ask for the politically impossible. The federal bureaucracy lives by politics. The people who make decisions do so always with an awareness of what will happen—inside the agency, inside the bureaucracy, in Congress, among the clientele, in the press, among interested groups. Some things that ought to be done cannot be done at the time because they are politically impossible. The subject has not come into its time. One has to work at politics and propaganda to build the acceptance and support that must join with change. Only the naive will ask for impossible change and expect to get it.

Similarly, only the naive mutter despair at the inadequacies of bureaucracy. The successful take advantage of them to get what they want. If this makes them seem unloving, hard, insensitive, unfeeling, others call their attitude realistic. They can hardly hope to change the federal bureaucracy in one lifetime. They want to use it for the good of society. They accept it on its own terms and take advantage of its weakness.

The bureaucracy is inefficient. The wise individual knows how to avoid the traps of inefficiency, such matters as useless copies and conferences, that make up the world of inefficiency. He also knows that to be efficient in the midst of inefficiency is to succeed like the moon on a clear autumn night.

The bureaucracy is bent on survival. Many of its people think first of how an action will affect the reputation and security of the agency. Will it gain or lose in relation with other agencies? Will the individuals who think this way gain or lose? The wise seeker of change will make his proposals in a frame that will make the agency and its officials gain in standing and survival.

The bureaucracy is uncoordinated. A successful advocate will find the loose parts, the vulnerable holes within bigness, and he and his kind will know how to use confusion or vacuum, whichever they find.

Finally, the successful politician who is working change through the bureaucracy will be patient. He does not tire after one attempt. He does not feel injured by having to go from one office to another until he finds the place where something will happen. He knows that in anything as big and sprawling as the federal bureaucracy, he must spend a lot of time and effort to succeed, and he will be patient while spending it.

The savers and protectors of the environment steadily use politics and propaganda, government and business, the professions, and bureaucracy to get changes. Never does change come easily or fast. But it comes all the time. If we had to do the impossible and now in the late years of the twentieth century say what chance life has of survival in the nation and the earth, we would say about an even chance. The physical-technological part of survival is much better than even. Success in using the social instruments is more doubtful, and this makes the average hope about even.

N O T E S

1. Jonathan Daniels, *Frontier on the Potomac* (Crowell-Collier and Macmillan, New York, 1946), pp. 31–32.
2. Robert Townsend, *Up the Organization, How to Stop the Organization from Stifling People and Strangling Profits* (Alfred A. Knopf, New York, 1970), p. 38.
3. *Ibid.*, p. 36.
4. David Halberstam, "The Programming of Robert McNamara," *Harper's Magazine*, February 1971, p. 60.
5. An acronym used among computer people and others. It stands for Garbage In, Garbage Out.

INDEX

Index

Printed in the United States
by Baker & Taylor Publisher Services